Ben Stacy Jerrik (Ed.)

Dąb, Greater Poland Voivodeship

AF708117

Ben Stacy Jerrik (Ed.)

Dąb, Greater Poland Voivodeship

Gmina Szamocin, Village, Gmina, Chodzież County, Greater Poland Voivodeship

Part Press

Publisher:
Part Press is a trademark of
International Book Market Service Ltd., 17 Rue Meldrum, Beau Bassin, 1713-01 Mauritius
Email: info@bookmarketservice.com
Website: www.bookmarketservice.com

Published in 2013

Printed in: U.S.A., U.K., Germany. This book was not produced in Mauritius.

ISBN: 978-613-8-91414-3

Contents

Articles

References

Dąb,_Greater_Poland_Voivodeship

Dąb

— Village —

Dąb

Coordinates: 52°25′20″N 18°20′5″E

Country	Poland
Voivodeship	Greater Poland
County	Konin County
Gmina	Skulsk

Dąb [dɔmp] is a village in the administrative district of Gmina Skulsk, within Konin County, Greater Poland Voivodeship, in west-central Poland.[1] It lies approximately 7 kilometres (4 mi) south of Skulsk, 24 km (15 mi) north of Konin, and 97 km (60 mi) east of the regional capital Poznań.

References

[1] "Central Statistical Office (GUS) - TERYT (National Register of Territorial Land Apportionment Journal)" (http://www.stat.gov.pl/broker/access/prefile/listPreFiles.jspa) (in Polish). 2008-06-01. .

Village

A **village** is a clustered human settlement or community, larger than a hamlet with the population ranging from a few hundred to a few thousand (sometimes tens of thousands). Though often located in rural areas, the term urban village is also applied to certain urban neighbourhoods, such as the East Village in Manhattan, New York City and the Saifi Village in Beirut, Lebanon, as well as Hampstead Village in the London conurbation. Villages are normally permanent, with fixed dwellings; however, transient villages can occur. Further, the dwellings of a village are fairly close to one another, not scattered broadly over the landscape, as a dispersed settlement.

The village of Rougon in Provence, France

In the past, villages were a usual form of community for societies that practise subsistence agriculture, and also for some non-agricultural societies. In Great Britain, a hamlet earned the right to be called a village when it built a church.[1] In many cultures, towns and cities were few, with only a small proportion of the population living in them. The Industrial Revolution attracted people in larger numbers to work in mills and factories; the concentration of people caused many villages to grow into towns and cities. This also enabled specialization of labor and crafts, and development of many trades. The trend of urbanization continues, though not always in connection with industrialisation. Villages have been eclipsed in importance as units of human society and settlement.

An alpine village in the Lötschental Valley, Switzerland

Traditional villages

Although many patterns of village life have existed, the typical village was small, consisting of perhaps 5 to 30 families. Homes were situated together for sociability and defence, and land surrounding the living quarters was farmed. Traditional fishing villages were based on artisan fishing and located adjacent to fishing grounds.

Berber village in Ourika valley, High Atlas, Morocco.

South Asia

India "The soul of India lives in its villages", declared M. K. Gandhi[2] at the beginning of 20th century. According to the 2011 census of India, 68.84% of Indians (around 833.1 million people) live in 640,867 different villages.[3] The size of these villages varies considerably. 236,004 Indian villages have a population less than 500, while 3,976 villages have a population of 10,000+. Most of the villages have their own temple, mosque or church depending on the local religious following.

A North Indian village in Rajasthan, India

A South Indian Village in Andhra Pradesh, India

East Asia

People's Republic of China In mainland China, villages are divisions under township Zh:or town Zh:.

Shirakawa-gō, Gifu Japan.

Republic of China (Taiwan) In the Republic of China (Taiwan), villages are divisions under townships or county-controlled cities. The village is called a *tsuen* or *cūn* () under a rural township () and a *li* () under an urban township () or a county-controlled city. See also Li (unit).

Japan

South Korea

Southeast Asia

Thailand

Brunei, Indonesia, Malaysia and Singapore

The *nagari* of Pariangan, West Sumatra.

In Indonesia, depending on the principles they are administered, villages are called *desa* or *kelurahan*. A *desa* (a term that derives from a Sanskrit word meaning "country" that is found in a name such as "Bangladesh") is administered according to traditions and customary law (*adat*), while a *kelurahan* is administered along more "modern" principles. *Desa* are generally located in rural areas while *kelurahan* are generally urban subdivisions. A village head is respectively called *kepala desa* or *lurah*. Both are elected by the local community. A *desa* or *kelurahan* is itself the subdivision of a *kecamatan* (district), in turn the subdivision of a *kabupaten* (regency).

The same general concept applies all over Indonesia. How ever, there is some variation among the vast numbers of Austronesian ethnic groups. For instance, in Bali villages have been created by grouping traditional hamlets or *banjar*, which constitute the basis of Balinese social life. In the Minangkabau country in West Sumatra province traditional villages are called *nagari* (a term deriving from another Sanskrit word meaning "city", which can be found in a name like "Srinagar"). In some areas such as Tanah Toraja, elders take turns watching over the village at a command post. As a general rule, *desa* and *kelurahan* are groupings of hamlets (*kampung* in Indonesian, *dusun* in the Javanese language, *banjar* in Bali).

In Malaysia, the term *kampung* (sometimes spelling *kampong*) in the English language has been defined specifically as "a Malay hamlet or village in a Malay-speaking country".[4] In other words, a ***kampung*** is defined today as a village in Brunei, Indonesia, Singapore, and Malaysia. In Malaysia, a *kampung* is determined as a locality with 10,000 or fewer people. Since historical times, every Malay village came under the leadership of a *penghulu* (village chief), who has the power to hear civil matters in his village (see Courts of Malaysia for more details). A Malay village typically contains a *"masjid"* (mosque) or *"surau"* (Muslim chapel), paddy fields and Malay houses on stilts. Malay and Indonesian villagers practice the culture of helping one another as a community, which is better known as "joint bearing of burdens" (*gotong royong*),[5] as well as being family-oriented (especially the concept of respecting one's family [particularly the parents and elders]), courtesy and believing in God (*"Tuhan"*) as paramount to everything else. It is common to see a cemetery near the mosque, as all Muslims in the Malay or Indonesian village want to be prayed for, and to receive Allah's blessings in the afterlife. While in Sarawak and East Kalimantan, some villages are called 'long', primarily inhabited by the Orang Ulu.

Singapore also follows the Malaysian *kampung*. However, there are only a few *kampung* villages remaining, mostly on islands surrounding Singapore such as Pulau Ubin. In the past, there were many *kampung* villages in Singapore but now there aren't many on the mainland.

The term "kampung", sometimes spelled "kampong" is one of many Malay words to have entered common usage in Malaysia and Singapore. Locally, the term is frequently used to refer to one's hometown.

Philippines

In urban areas of the Philippines, the term "village" most commonly refers to private subdivisions, especially gated communities. These villages emerged in the mid-20th century and were initially the domain of elite urban dwellers. Those are common in major cities in the country and their residents have a wide range of income levels. Such villages may or may not correspond to administrative units (usually barangays) and/or be privately administered. Barangays more correspond to the villages of old times, and the chairman (formerly a village datu) now settles administrative, intrapersonal, and political matters or polices the village, though with much less authority and respect than in Indonesia or Malaysia.

Vietnam

Village, or "làng", is a basis of Vietnam society. Vietnam's village is the typical symbol of Asian agricultural production. Vietnam's village typically contains: a village gate, "lũy tre" (bamboo hedges), "đình làng" (communal house) where "thành hoàng" (tutelary god) is worshiped, a common well, "đồng lúa" (rice field), "chùa" (temple) and houses of all families in the village. All the people in Vietnam's villages usually have a blood relationship. They are farmers who grow rice and have the same traditional handicraft. Vietnam's villages have an important role in society (Vietnamese saying: "Custom rules the law" -"Phép vua thua lệ làng" [literally: the king's law yields to village customs]). Everyone in Vietnam wants to be buried in their village when they die.

Central and Eastern Europe

Slavic countries

Selo (Cyrillic: село; Polish: *wieś, sioło*) is a Slavic word meaning "village" in Bosnia and Herzegovina, Bulgaria, Croatia, Macedonia, Russia, Serbia, and Ukraine. For example there are numerous *sela* (plural of *selo*) called Novo Selo in Bulgaria, Croatia, Montenegro and others in Serbia, and Macedonia. In Slovenia, the word *selo* is used for very small villages (less than a thousand people) and in dialects; the Slovene word *vas* is used all over Slovenia.

Bulgaria

Kovachevitsa, a village in southern Bulgaria

In Bulgaria, the different types of *Sela* vary from a small selo of 5 to 30 families to one of several thousand people. According to a 2002 census, in that year there were 2,385,000 Bulgarian citizens living in settlements classified as *villages*.[6] A 2004 Human Settlement Profile on Bulgaria[7] conducted by the United Nations Department of Economic and Social Affairs stated that:

> The most intensive is the migration "city – city". Approximately 46% of all migrated people have changed their residence from one city to another. The share of the migration processes "village – city" is significantly less – 23% and "city – village" – 20%. The migration "village – village" in 2002 is 11%.[6]

It also stated that

> the state of the environment in the small towns and villages is good apart from the low level of infrastructure.[6]

In Bulgaria, it is becoming popular to visit villages for the atmosphere, culture, crafts, hospitality of the people and the surrounding nature. This is called *selski tourism* (Bulgarian: селски туризъм), meaning "village tourism".

Russia

The village near Lake Baikal, Siberia

In Russia, as of the 2010 Census, 26.3% of the country's population lives in rural localities;[8] down from 26.7% recorded in the 2002 Census.[8] Multiple types of rural localities exist, but the two most common are *derevnya* (деревня) and *selo* (село). Historically, the formal indication of status was religious: a city (*gorod*) had a cathedral, a *selo* had a church, while a *derevnya* had neither.

The lowest administrative unit of the Russian Empire, a *volost*, or its Soviet or modern Russian successor, a *selsoviet*, was typically headquartered in a *selo* and embraced a few neighboring villages.

Between 1926 and 1989, Russia's rural population shrank from 76 million people to 39 million, due to urbanization, collectivization, dekulakization, and the World War II losses, but has nearly stabilized since. During 1930–1937, mass starvation in Russia and other parts of the Soviet Union lead to the death of at least 14.5 million peasants (including 5–7 million in the Holodomor).[9]

Most Russian rural localities have populations of less than 200 people, and the smaller places take the brunt of depopulation: e.g., in 1959, about one half of Russia's rural population lived in villages of fewer than 500 people, while now less than one third does. In the 1960s–1970s, the depopulation of the smaller villages was driven by the central planners' drive to get the farm workers out of smaller, "prospect-less" hamlets and into the collective or state farms' main villages, with more amenities.[10]

Most Russian rural residents are involved in agricultural work, and it is very common for villagers to produce their own food. As prosperous urbanites purchase village houses for their second homes, Russian villages sometimes are transformed into dacha settlements, used mostly for seasonal residence.

The historically Cossack regions of Southern Russia and parts of Ukraine, with their fertile soil and absence of serfdom, had a rather different pattern of settlement from central and northern Russia. While peasants of central Russia lived in a village around the lord's manor, a Cossack family often lived on its own farm, called *khutor*. A number of such *khutors* plus a central village made up the administrative unit with a center in a *stanitsa* (Russian: стани́ца; Ukrainian: станиця, *stanytsia*). Such *stanitsas* often with a few thousand residents, were usually larger than a typical *selo* in central Russia.

The term *aul/aal* is used to refer mostly Muslim-populated villages in Caucasus and Idel-Ural, without regard to the number of residents.

Ukraine

In Ukraine, a village, known locally as a *"selo"* (село), is considered the lowest administrative unit. Villages may have an individual administration (*silrada*) or a joint administration, combining two or more villages. Villages may also be under the jurisdiction of a city council (*miskrada*) or town council (*selyshchna rada*) administration.

There is, however, another smaller type of settlement which is designated in Ukrainian as a *selysche* (селище). This type of community is generally referred to in English as a "settlement". In comparison with an urban-type settlement, Ukrainian legislation does not have a concrete definition or a criterion to differentiate such settlements from villages. They represent a type of a small rural locality that might have once been a *khutir*, a fisherman's settlement, or a dacha. They are administered by a *silrada* (council) located in a nearby adjacent village. Sometimes the term *"selysche"* is also used in a more general way to refer to adjacent settlements near a bigger city, including urban-type settlements (*selysche miskoho typu*) and/or villages; however, ambiguity is often avoided in connection with urbanized settlements by referring to them using the three-letter abbreviation *smt* instead.

The largest Ukrainian village ("selo") Kosmach

The *khutir* (хутір) and *stanytsia* (станиця) are not part of the administrative division any longer, primarily due to collectivization. *Khutirs* were very small rural localities consisting of just few housing units and were sort of individual farms. They became really popular during the Stolypin reform in the early 20th century. During the collectivization, however, residents of such settlements were usually declared to be kulaks and had all their property confiscated and distributed to others (nationalized) without any compensation. The *stanitsa* likewise has not survived as an administrative term. The *stanitsa* was a type of a collective community that could include one or more settlements such as villages, *khutirs*, and others. Today,

stanitsa-type formations have only survived in Kuban (Russian Federation) where Ukrainians were resettled during the time of the Russian Empire.

Western & Southern Europe

United Kingdom

A village in the UK is a compact settlement of houses, smaller in size than a town, and generally based on agriculture or, in some areas, mining (such as Ouston, County Durham), quarrying or sea fishing.

The major factors in the type of settlement are location of water sources, organisation of agriculture and landholding, and likelihood of flooding. For example, in areas such as the Lincolnshire Wolds, the villages are often found along the spring line halfway down the hillsides, and originate as spring line settlements, with the original open field systems around the village. In northern Scotland, most villages are planned to a grid pattern located on or close to major roads, whereas in areas such as the Forest of Arden, woodland clearances produced small hamlets around village greens.[11] [12]

The main street of the village of Castle Combe, Wiltshire, England.

Some villages have disappeared (for example, deserted medieval villages), sometimes leaving behind a church or manor house and sometimes nothing but bumps in the fields.Some show archaeological evidence of settlement at three or four different layers, each distinct from the previous one. Clearances may have been to accommodate sheep or game estates, or enclosure, or may have resulted from depopulation, such as after the Black Death or following a move of the inhabitants to more prosperous districts. Other villages have grown and merged and often form hubs within the general mass of suburbia — such as Hampstead, London and Didsbury in Manchester. Many villages are now predominantly dormitory locations and have suffered the loss of shops, churches and other facilities.

For many British people, the village represents an ideal of Great Britain. Seen as being far from the bustle of modern life, it is represented as quiet and harmonious, if a little inward-looking. This concept of an unspoilt Arcadia is present in many popular representations of the village such as the radio serial *The Archers* or the best kept village competitions.[13]

Bisley, Gloucestershire, a village in the Cotswolds

Many villages in South Yorkshire, North Nottinghamshire, North East Derbyshire, County Durham, South Wales and Northumberland are known as pit villages. These (such as Murton, County Durham) grew from hamlets when the sinking of a colliery in the early 20th century resulted in a rapid growth in their population and the colliery owners built new housing, shops, pubs and churches. Some pit villages outgrew nearby towns by area and population; for example, Rossington in South Yorkshire came to have over four times more people than the nearby town of Bawtry. Some pit villages grew to become towns; for example, Maltby in South Yorkshire grew from 600 people in the 19th century[14] to over 17,000 in 2007.[15] Maltby was constructed under the auspices of the Sheepbridge Coal and Iron Company and included ample open spaces and provision for gardens.[16]

In the UK, the main historical distinction between a hamlet and a village was that the latter had a church,[1] and so usually was the centre of worship for an ecclesiastical parish. However, some civil parishes may contain more than one village. The typical village had a pub or inn, shops, and a blacksmith. But many of these facilities are now gone, and many villages are dormitories for commuters. The population of such settlements ranges from a few hundred people to around five thousand. A village is distinguished from a town in that:

- A village should not have a regular agricultural market, although today such markets are uncommon even in settlements which clearly are towns.
- A village does not have a town hall nor a mayor.
- If a village is the principal settlement of a civil parish, then any administrative body that administers it at parish level should be called a parish council or parish meeting, and not a town council or city council. However, some civil parishes have no functioning parish, town, or city council nor a functioning parish meeting. In Wales, where the equivalent of an English civil parish is called a Community, the body that administers it is called a Community Council. However, larger councils may elect to call themselves town councils.[17] Unlike Wales, Scottish community councils have no statutory powers.[18]
- There should be a clear green belt or open fields, as, for example, seen on aerial maps for Ouston surrounding its parish[19] borders. However this may not be applicable to urbanised villages: although these may not considered to be villages, they are often widely referred to as being so; an example of this is Horsforth in Leeds.

France

Same general definition as in the UK.

An independent association named *Les Plus Beaux Villages de France*, was created in 1982 to promote assets of small and picturesque French villages of quality heritage. As of 2008, 152 villages in France have been listed in "The Most Beautiful Villages of France".

Saint-Cirq-Lapopie (Lot) is one of "The Most Beautiful Villages in France".

Spain

Spain has plenty of little villages around its territory. The concept of village and country life is really present and usual in the North of the country (Atlantic area), especially in Galicia where villages are similar to English ones.

South of Barcelona is Spain's most romantic Mediterranean beach town, with a 2.5 km-long (1 1/2-mile) sandy beach and a promenade studded with flowers and palm trees. Sitges is a town with a rich connection to art; Picasso and Dalí both spent time here.[20] Mérida is an important Roman town with great tapas. Barcena Mayor (Cantabria) has houses that date back to the sixth century with simple two floors constructions and rectangular form. Salamanca, an ancient Celtic town, is also a Renaissance city with striking architecture. Its sandstones buildings have a beautiful lustre giving the city the nickname, La Ciudad Dorado.[21]

Morella, Castellón is a medieval village located in the region of "Comunitat Valenciana" with huge castles with a rich renaissance history.[22] Rogueira pasturelands is one of the great ecological jewels of Galicia. Rivers, pools and springs abound in this verdant forest, as do underground water caves and caverns with a prehistoric past.[23] San Marti Vell is a charming small village well known for its Gothic spire. La Bisbal should be next on the list. The town is worth visiting for its Main Square and the castle. This Romanesque castle is situated in the middle of the town, giving it a romantic look. There is also Palafrugell, Palau-sator, Sant Julia and Sant Feliu de Boada. They are all very important because of their medieval patrimony. Castelló d'Empúries has 13th century Gothic churches. Angles also possesses outstanding medieval constructions throughout its village.[24] All villages have a church or hermitage.

Portugal

Villages are more usual in the northern and central regions and in the Alentejo. Most of them have a church and a "Casa do Povo" (people's house), where the village's summer **romarias** or religious festivities are usually held. Summer is also when many villages are host to a range of folk festivals and fairs, taking advantage of the fact that many of the locals who reside abroad tend to come back to their native village for the holidays.

Netherlands

In the flood prone districts of the Netherlands, villages were traditionally built on low man-made hills called terps before the introduction of regional dyke-systems. In modern days, the term *dorp* (lit. "village") is usually applied to settlements no larger than 20,000, though there's no official law regarding status of settlements in the Netherlands.

Middle East

Lebanon

The main square of Saifi Village in Centre Ville, Beirut, Lebanon

Like France, villages in Lebanon are usually located in remote mountainous areas. The majority of villages in Lebanon retain their Aramaic names or are derivative of the Aramaic names, and this is because Aramaic was still in use in Mount Lebanon up to the 18th century.[25]

Many of the Lebanese villages are a part of districts, these districts are known as "kadaa" which includes the districts of Baabda (Baabda), Aley (Aley), Matn (Jdeideh), Keserwan (Jounieh), Chouf (Beiteddine), Jbeil (Byblos), Tripoli (Tripoli), Zgharta (Zgharta / Ehden), Bsharri (Bsharri), Batroun (Batroun), Koura (Amioun), Miniyeh-Danniyeh (Minyeh / Sir Ed-Danniyeh), Zahle (Zahle), Rashaya (Rashaya), Western Beqaa (Jebjennine / Saghbine), Sidon (Sidon), Jezzine (Jezzine), Tyre (Tyre), Nabatiyeh (Nabatiyeh), Marjeyoun (Marjeyoun), Hasbaya (Hasbaya), Bint Jbeil (Bint Jbeil), Baalbek (Baalbek), and Hermel (Hermel).

The district of Danniyeh consists of thirty six small villages, which includes Almrah, Kfirchlan, Kfirhbab, Hakel al Azimah, Siir, Bakhoun, Miryata, Assoun, Sfiiri, Kharnoub, Katteen, Kfirhabou, Zghartegrein, Ein Qibil.

Danniyeh (known also as Addinniyeh, Al Dinniyeh, Al Danniyeh, Arabic: سير الضنية) is a region located in Miniyeh-Danniyeh District in the North Governorate of Lebanon. The region lies east of Tripoli, extends north as far as Akkar District, south to Bsharri District and Zgharta District and as far east as Baalbek and Hermel. Dinniyeh has an excellent ecological environment filled with woodlands, orchards and groves. Several villages are located in this mountainous area, the largest town being Sir Al Dinniyeh.

An example of a typical mountainous Lebanese village in Dannieh would be Hakel al Azimah which is a small village that belongs to the district of Danniyeh, situated between Bakhoun and Assoun's boundaries. It is in the centre of the valleys that lie between the Arbeen Mountains and the Khanzouh.

Syria

General view from Al-Annaze village, near Tartus, Syria

Syria contains a large number of villages that vary in size and importance, including the ancient, historical and religious villages, such as Ma'loula, Sednaya, and Brad (Mar Maroun's time). The diversity of the Syrian environments creates significant differences between the Syrian villages in terms of the economic activity and the method of adoption. Villages in the south of Syria (Hauran, Jabal al-Druze), the north-east (the Syrian island) and the Orontes River basin depend mostly on agriculture, mainly grain, vegetables and fruits. Villages in the region of Damascus and Aleppo depend on trading. Some other villages, such as Marmarita depend heavily on tourist activity.

Mediterranean cities in Syria, such as Tartus and Latakia have similar types of villages. Mainly, villages were built in very good sites which had the fundamentals of the rural life, like water. An example of a Mediterranean Syrian village in Tartus would be al-Annaze, which is a small village that belongs to the area of al-Sauda. The area of al-Sauda is called a nahiya, which is a subdistrict.

Australasia and Oceania

The village of Puamau on Hiva Oa, Marquesas Islands, French Polynesia

Pacific Islands Communities on pacific islands were historically called villages by English speakers who traveled and settled in the area. Some communities such as several Villages of Guam continue to be called villages despite having large populations that can exceed 40,000 residents.

New Zealand The traditional Māori village was the pā, a fortified hill-top settlement. Tree-fern logs and flax were the main building materials. As in Australia (see below) the term is now used mainly in respect of shopping or other planned areas.

Australia The term village often is used in reference to small planned communities such as retirement communities or shopping districts, and tourist areas such as ski resorts. Small rural communities are usually known as townships. Larger settlements are known as towns.

South America

Argentina Usually set in remote mountainous areas, some also cater to winter sports and/or tourism, see: Uspallata, La Cumbrecita, Villa Traful and La Cumbre

North America

Canada

A Newfoundland fishing village

United States

Incorporated villages

A church in Newfane, Vermont

In twenty[26] U.S. states, the term "village" refers to a specific form of incorporated municipal government, similar to a city but with less authority and geographic scope. However, this is a generality; in many states, there are villages that are an order of magnitude larger than the smallest cities in the state. The distinction is not necessarily based on population, but on the relative powers granted to the different types of municipalities and correspondingly, different obligations to provide specific services to residents.

In some states such as New York, Wisconsin, or Michigan, a village is an incorporated municipality, usually, but not always, within a single town or civil township. Residents pay taxes to the village and town or township and may vote in elections for both as well. In some cases, the village may be coterminous with the town or township. There are also many villages which span the boundaries of more than one town or township, and some villages may even straddle county borders.

There is no limit to the population of a village in New York; Hempstead, the largest village in the state, has 55,000 residents, making it more populous than some of the state's cities. However, villages in the state may not exceed five square miles (13 km²) in area.

In the state of Wisconsin, a village is always legally separate from the towns that it has been incorporated from. The largest village is Menomonee Falls, which has over 32,000 residents.

Michigan and Illinois also have no set population limit for villages and there are many villages that are larger than cities in those states. The village of Arlington Heights, IL had 75,101 residents as of the 2010 census.

Villages in Ohio are often legally part of the township from which they were incorporated, although exceptions such as Hiram exist, in which the village is separate from the township.[27] They have no area limitations, but become cities if they grow a population of more than 5,000.[28]

In Maryland, a locality designated "Village of ..." may be either an incorporated town or a special tax district.[29] An example of the latter is the Village of Friendship Heights.

In states that have New England towns, a "village" is a center of population or trade, including the town center, in an otherwise sparsely-developed town or city — for instance, the village of Hyannis in the city of the Barnstable, Massachusetts.

Unincorporated villages

In many states, the term "village" is used to refer to a relatively small unincorporated community, similar to a hamlet in New York state. This informal usage may be found even in states that have villages as an incorporated municipality, although such usage might be considered incorrect and confusing.

Africa

Nigeria

Villages in Nigeria vary significantly because of cultural and geographical differences.

Northern Nigeria

In the North, villages were under traditional rulers long before the Jihad of Shaikh Uthman Bin Fodio and after the Holy War. At that time Traditional rulers used to have absolute power in their administrative regions. After Dan Fodio's Jihad in 1804,[30] political structure of the North became Islamic where emirs were the political, administrative and spiritual leaders of their people. These emirs appointed a number of people to assisted them in running the administration and that included villages. [31]

A village in Kaita, NigeriaKaita Nigeria

Every Hausa village was reigned by Magaji (Village head) who was answerable to his Hakimi (mayor) at town level. The Magaji also had his cabinet who assisted him rule his village efficiently, among whom was Mai-Unguwa (Ward Head). [32]

With the creation of Native Authority in Nigerian provinces, the autocratic power of village heads along with all other traditional rulers was subdued hence they ruled 'under the guidance of colonial officials'.[33]

Even though the constitution of the Federal Republic of Nigeria has not recognised the functions of traditional rulers, they still command respect in their villages[33] and political office holders liaise with them almost every time to reach people.

In Hausa language, village is called **ƙauye** and every local government area is made up of several small and large **ƙauyuka** (villages). For instance, Girka is a village in Kaita town in Katsina state in Nigeria. They have mud houses with thatched roofing though, like in most of villages in the North, zinc roofing is becoming a common sight.

Still in many villages in the North, people do not have access to portable water.[34] So they fetch water from ponds and streams. Others are lucky to have wells within a walking distance. Women rush in the morning to fetch water in their clay pots from wells, boreholes and streams. However, government is now providing them with water bore holes.[35]

Electricity and GSM network are reaching more and more villages in the North almost everyday. So bad feeder roads may lead to remote villages with electricity and unstable GSM network.[36]

Southern Nigeria

Village dwellers in the Southeastern region lived separately in 'clusters of huts belonging to the patrilinage'.[37] As the rainforest region is dominated by Igbo speaking people, the villages are called **ime obodo** (inside town) in Igbo language. A typical large village might have a few thousand persons who shared the same market, meeting place and beliefs.

See also

- Global village
- Linear village
- Village green
- Village lock-up
- police village

Settlement types

- Dugout
- Fishing village
- Hamlet
- Microtown

Countries and localities

- Dhani and villages
- Dogon villages
- Hakka architecture
- Ksar
- List of villages in Europe by country
- Pueblo
- Sołectwo (rough equivalent in Poland)
- Ville

Developed environments

- Developed environments
- City
- Exurban
- Megalopolis
- Rural
- Suburban
- Urban area

Footnotes

[1] Dr Greg Stevenson, "What is a Village?" (http://www.bbc.co.uk/history/programmes/restoration/2006/exploring_brit_villages_01.shtml), *Exploring British Villages*, BBC, 2006, accessed 20 October 2009

[2] R.K. Bhatnagar. INDIA'S MEMBERSHIP OF ITER PROJECT (http://www.pibbng.kar.nic.in/feature1.pdf). PRESS INFORMATION BUREAU. GOVERNMENT OF INDIA, BANGALORE

[3] "Indian Census" (http://www.censusindia.gov.in/). Censusindia.gov.in. . Retrieved 2012-04-09.

[4] "Meriam-Webster Online" (http://www.m-w.com/dictionary/kampung). M-w.com. 2007-04-25. . Retrieved 2010-03-28.

[5] Geertz, Clifford. "Local Knowledge: Fact and Law in Comparative Perspective", pp. 167–234 in Geertz *Local Knowledge: Further Essays in Interpretive Anthropology,* NY: Basic Books. 1983.

[6] "Human Settlement Country Profile, Bulgaria (*2004*)" (http://www.un.org/esa/agenda21/natlinfo/countr/bulgaria/Bulgariahumansettlement2003.PDF) (PDF). United Nations Department of Economic and Social Affairs. . Retrieved 2008-11-30.

[7] HUMAN SETTLEMENT COUNTRY PROFILE: BULGARIA (http://www.un.org/esa/agenda21/natlinfo/countr/bulgaria/Bulgariahumansettlement2003.PDF). United Nations (2004)

[8] "Предварительные итоги Всероссийской переписи населения 2010 года[[Category:Articles containing non-English language text (http://www.perepis-2010.ru/results_of_the_census/results-inform.php)] [Preliminary results of the 2010 All-Russian Population Census]"] (in

Russian). *Всероссийская перепись населения 2010 года (2010 All-Russia Population Census)*. Federal State Statistics Service. 2011. . Retrieved February 9, 2012.

[9] Robert Conquest (1986) *The Harvest of Sorrow: Soviet Collectivization and the Terror-Famine*. Oxford University Press. ISBN 0-19-505180-7.

[10] "Российское село в демографическом измерении" (*Rural Russia measured demographically*) (http://demoscope.ru/weekly/2006/0253/tema04.php) (**Russian**). This article reports the following census statistics:

Census year	1959	1970	1979	1989	2002
Total number of rural localities in Russia	294,059	216,845	177,047	152,922	155,289
Of them, with population 1 to 10 persons	41,493	25,895	23,855	30,170	47,089
Of them, with population 11 to 200 persons	186,437	132,515	105,112	80,663	68,807

[11] Wild, Martin Trevor (2004). *Village England: a social history of the countryside* (http://books.google.co.uk/books?id=M7AmyuGr5Y8C&printsec=frontcover). I.B.Tauris. p. 12. ISBN 978-1-86064-939-4. .

[12] Taylor, Christopher (1984). *Village and farmstead: A history of rural settlement in England* (http://books.google.co.uk/books?ei=NksHTpXqIo6t8QPE1ozADQ). G. Philip. p. 192. ISBN 978-0-540-01082-0. .

[13] OECD (2011). *OECD Rural Policy Reviews: England, United Kingdom 2011* (http://books.google.co.uk/books?id=bQCGWKfXJNMC&printsec=frontcover&source=gbs_ge_summary_r&cad=0). OECD Publishing. p. 237. .

[14] *The Parliamentary gazetteer of England and Wales* (http://books.google.co.uk/books?id=mxIQAAAAYAAJ&printsec=frontcover). **3**. A. Fullarton & Co.. 1851. p. 344. .

[15] "Maltby Ward" (http://www.rotherham.gov.uk/download/553/maltby_ward). Rotherham Metropolitan Borough Council. . Retrieved 2011-06-26.

[16] Baylies, Carolyn Louise (1993). *The history of the Yorkshire miners, 1881–1918* (http://books.google.co.uk/books?id=WEIOAAAAQAAJ&printsec=frontcover&source=gbs_ge_summary_r&cad=0). Routledge. .

[17] "National Statistics" (http://www.statistics.gov.uk/geography/parishes.asp). Statistics.gov.uk. . Retrieved 2010-03-28.

[18] "Portobello Community Council" (http://www.porty.org.uk/council/index.php). Porty.org.uk. . Retrieved 2010-03-28.

[19] "Ouston Parish Council" (http://parishes.durham.gov.uk/ouston/Pages/wherewelive.aspx). durham.gov.uk. .

[20] The Best Small Towns and Villages in Spain at Frommer's (http://www.frommers.com/destinations/spain/0242020855.html). Frommers.com. Retrieved on 2012-05-27.

[21] Spanish Towns and Villages (http://www.tourclare.com/spanishtownsandvillages.php). Tourclare.com. Retrieved on 2012-05-27.

[22] Top 10 Cities and Villages to Visit in Spain (http://www.travelthruspain.com/what-to-do/cities). Travelthruspain.com. Retrieved on 2012-05-27.

[23] Medieval villages in Galicia, Spain: O courel mountains. spain.info in English (http://www.spain.info/no/reportajes/sierra_de_o_courel_naturaleza_y_aldeas_medievales.html). Spain.info. Retrieved on 2012-05-27.

[24] Medieval Towns And Villages In Spain (http://travel.ezinemark.com/medieval-towns-and-villages-in-spain-16bc87fb4dc.html). Travel.ezinemark.com (2010-10-22). Retrieved on 2012-05-27.

[25] "A project proposal" (http://almashriq.hiof.no/lebanon/400/410/412/elies_project/glimse_of_yesterday.html). Almashriq.hiof.no. . Retrieved 2010-03-28.

[26] "Village" (http://www.websters-online-dictionary.org/definition/english/vi/village.html#Definitions). Websters-online-dictionary.org. . Retrieved 2010-03-28.

[27] "Detailed map of Ohio" (http://www2.census.gov/geo/maps/general_ref/cousub_outline/cen2k_pgsz/oh_cosub.pdf) (PDF). United States Census Bureau. 2000. . Retrieved 2010-03-28.

[28] "Ohio Revised Code Section 703.01(A)" (http://codes.ohio.gov/orc/703.01). . Retrieved 2010-03-28.

[29] 2002 Census of Governments, Individual State Descriptions (http://www.census.gov/prod/2005pubs/gc021x2.pdf) (PDF)

[30] Britannica Encylopedia **The New Encyclopaedia Britannica** Vol. 6, 15th Edition. isbn 0-85229-961-3. Page 763

[31] Dr. Sani Abubakar Lugga. **The Great Province** published by Lugga Press Gidan Lugga, Kofar Marusa Road, Katsina Nigeria. ISBN 978-2105- 48 - 1. Page 43

[32] Dr. Sani Abubakar Lugga. **The Great Province** published by Lugga Press Gidan Lugga, Kofar Marusa Road, Katsina Nigeria. ISBN 978-2105- 48 - 1. Page 63

[33] A Johnson Ugoji Anyaele. **Comprehensive Government** A Johnson Publishers LTD. Surulere, Lagos. ISBN 978-2799-49-1. Page 123

[34] Studies on well water and possible health risks in Katsina, Nigeria. http://www.ncbi.nlm.nih.gov/pmc/articles/PMC2134251/

[35] Katsina state government spends N4.2 billion on water supply project each year. http://www.peoplesdaily-online.com/index.php/news/news/special-report/7632-how-katsina-state-is-doing-so-much-with-so-little

[36] Nigerian Operator Expands Coverage. http://www.cellular-news.com/story/16837.php

[37] Villages in Igbo land prior to the colonial era. http://www.igboguide.org/HT-chapter10.htm

External links

- Types of villages (anthropogenic biomes) (http://www.ecotope.org/anthromes/v1/guide/villages/)

Gmina_Krzymów

Gmina Krzymów

Krzymów Commune

— Gmina —

Coat of arms

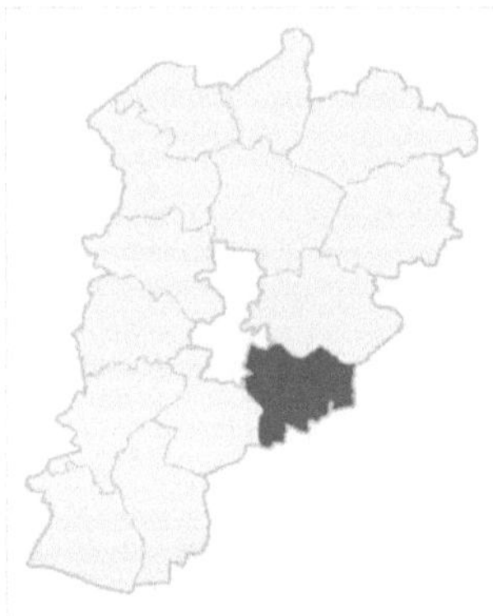

Coordinates (Krzymów): 52°12′N 18°26′E

Country	Poland
Voivodeship	Greater Poland
County	Konin County
Seat	Krzymów
Area	
• Total	92.68 km^2 (**unknown operator: u'strong'** sq mi)
Population (2006)	
• Total	7074
• Density	**unknown operator: u'strong'**/km^2 (**unknown operator: u'strong'**/sq mi)
Website	[1]

Gmina Krzymów is a rural gmina (administrative district) in Konin County, Greater Poland Voivodeship, in west-central Poland. Its seat is the village of Krzymów, which lies approximately 12 kilometres (7 mi) east of Konin and 106 km (66 mi) east of the regional capital Poznań.

The gmina covers an area of 92.68 square kilometres (35.8 sq mi), and as of 2006 its total population is 7,074.

Villages

Gmina Krzymów contains the villages and settlements of Adamów, Borowo, Brzezińskie Holendry, Brzeźno, Chójki, Drążeń, Drążno-Holendry, Genowefa, Głodno, Ignacew, Kałek, Krzymów, Nowe Paprockie Holendry, Nowy Krzymów, Paprotnia, Piersk, Potażniki, Rożek Brzeziński, Smolnik, Stare Paprockie Holendry, Szczepidło, Teresina and Zalesie.

Neighbouring gminas

Gmina Krzymów is bordered by the city of Konin and by the gminas of Kościelec, Kramsk, Stare Miasto, Tuliszków and Władysławów.

References

- Polish official population figures 2006 [2]

References

[1] http://www.krzymow.nowoczesnagmina.pl
[2] http://www.stat.gov.pl/gus/45_655_PLK_HTML.htm

Adamów,_Gmina_Krzymów

Adamów	
— Village —	
 Adamów	
Coordinates: 52°9′N 18°20′E	
Country	Poland
Voivodeship	Greater Poland
County	Konin County
Gmina	Krzymów

Adamów [aˈdamuf] is a village in the administrative district of Gmina Krzymów, within Konin County, Greater Poland Voivodeship, in west-central Poland.[1] It lies approximately 9 kilometres (6 mi) south-west of Krzymów (the gmina seat), 9 km (6 mi) south-east of Konin, and 101 km (63 mi) east of the regional capital Poznań.

References

[1] "Central Statistical Office (GUS) - TERYT (National Register of Territorial Land Apportionment Journal)" (http://www.stat.gov.pl/broker/access/prefile/listPreFiles.jspa) (in Polish). 2008-06-01. .

Borowo,_Konin_County

Borowo	
— Village —	
 Borowo	
Coordinates: 52°12′N 18°24′E	
Country	Poland
Voivodeship	Greater Poland
County	Konin County
Gmina	Krzymów

Borowo [bɔˈrɔvɔ] is a village in the administrative district of Gmina Krzymów, within Konin County, Greater Poland Voivodeship, in west-central Poland.[1] It lies 3 kilometres (2 mi) west of Krzymów, 10 km (6 mi) east of Konin, and 104 km (65 mi) east of the regional capital Poznań.

References

[1] "Central Statistical Office (GUS) - TERYT (National Register of Territorial Land Apportionment Journal)" (http://www.stat.gov.pl/broker/access/prefile/listPreFiles.jspa) (in Polish). 2008-06-01. .

Konin_County

Konin County **Powiat koniński**	
— County —	
 **Coat of arms**	
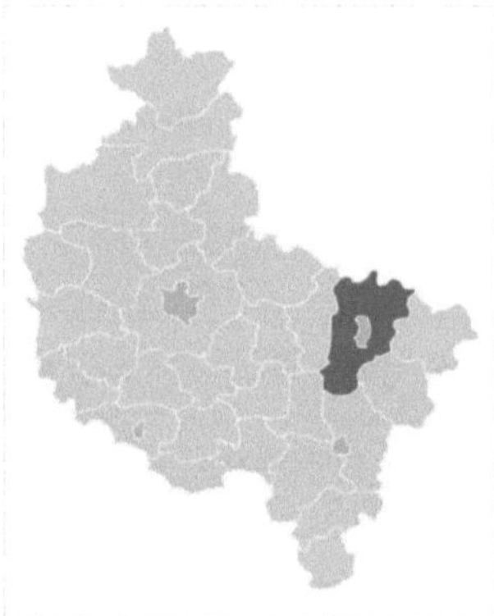 Location within the voivodeship	
Coordinates (Konin): 52°13′N 18°16′E	
Country	Poland
Voivodeship	Greater Poland
Seat	Konin
Gminas	
Area	
• **Total**	1578.71 km^2 (**unknown operator: u'strong'** sq mi)
Population (2006)	
• **Total**	123646
• **Density**	**unknown operator: u'strong'**/km^2 (**unknown operator: u'strong'**/sq mi)
• **Urban**	17677
• **Rural**	105969
Car plates	PKN

Konin County (Polish: *powiat koniński*) is a unit of territorial administration and local government (powiat) in Greater Poland Voivodeship, west-central Poland. It came into being on January 1, 1999, as a result of the Polish local government reforms passed in 1998. Its administrative seat is the city of Konin, although the city is not part of the county (it constitutes a separate city county). The county contains five towns: Golina, 12 km (7 mi) north-west of Konin, Kleczew, 19 km (12 mi) north of Konin, Sompolno, 27 km (17 mi) north-east of Konin, Ślesin, 17 km (11 mi) north of Konin, and Rychwał, 18 km (11 mi) south of Konin.

The county covers an area of 1578.71 square kilometres (609.5 sq mi). As of 2006 its total population is 123,646, out of which the population of Golina is 4,330, that of Kleczew is 4,173, that of Sompolno is 3,695, that of Ślesin is 3,102, that of Rychwał is 2,377, and the rural population is 105,969.

Neighbouring counties

Apart from the city of Konin, Konin County is also bordered by Mogilno County, Inowrocław County and Radziejów County to the north, Koło County to the east, Turek County to the south-east, Kalisz County to the south, Pleszew County to the south-west, and Słupca County to the west.

Administrative division

The county is subdivided into 14 gminas (five urban-rural and nine rural). These are listed in the following table, in descending order of population.

Gmina	Type	Area (km^2)	Population (2006)	Seat
Gmina Ślesin	urban-rural	145.7	13,446	Ślesin
Gmina Golina	urban-rural	99.1	11,326	Golina
Gmina Sompolno	urban-rural	137.4	10,530	Sompolno
Gmina Kazimierz Biskupi	rural	108.0	10,459	Kazimierz Biskupi
Gmina Stare Miasto	rural	97.8	10,355	Stare Miasto
Gmina Kramsk	rural	131.8	10,133	Kramsk
Gmina Kleczew	urban-rural	110.1	9,721	Kleczew
Gmina Rychwał	urban-rural	117.8	8,387	Rychwał
Gmina Wierzbinek	rural	148.1	7,597	Wierzbinek
Gmina Krzymów	rural	92.7	7,074	Krzymów
Gmina Rzgów	rural	104.7	6,818	Rzgów Pierwszy
Gmina Wilczyn	rural	83.1	6,422	Wilczyn
Gmina Skulsk	rural	84.9	6,145	Skulsk
Gmina Grodziec	rural	117.7	5,233	Grodziec

References

- Polish official population figures 2006 [2]

External links

- Official county website [1]
- Town website of Old City - www.konin-starowka.pl [2]

References

[1] http://www.powiat.konin.pl/
[2] http://www.konin-starowka.pl/

Greater_Poland_Voivodeship

Greater Poland Voivodeship
Województwo wielkopolskie

— Voivodeship —

wielkopolska
wielka historia
wielka przygoda

Flag

Coat of arms

Location within Poland

Division into counties

Coordinates (Poznań): 52°17′34″N 16°44′8″E	
Country	Poland
Capital	Poznań
Counties	
Area	
• **Total**	29826 km^2 (**unknown operator: u'strong'** sq mi)
Population (2006)	
• **Total**	3374653
• **Density**	**unknown operator: u'strong'**/km^2 (**unknown operator: u'strong'**/sq mi)
• **Urban**	1923582
• **Rural**	1451071
Car plates	P
Website	[1]
* further divided into 226 gminas.	

Wielkopolska Voivodeship or **Wielkopolska Province**[2] (in Polish, *województwo wielkopolskie* [vɔjɛˈvut͡stfɔ vjɛlkɔˈpɔlskʲɛ]), also known as **Greater Poland Voivodeship**, is a voivodeship, or province, in west-central Poland. It was created on 1 January 1999 out of the former Poznań, Kalisz, Konin, Piła and Leszno Voivodeships, pursuant to the Polish local government reforms adopted in 1998. The province is named after the region called Greater Poland or *Wielkopolska* [vjɛlkɔˈpɔlska] (listen). The modern province includes most of this historic region, except for some south-western parts.

Greater Poland Voivodeship is second in area and third in population among Poland's sixteen voivodeships, with an area of 29826 square kilometres (**unknown operator: u'strong'** sq mi) and a population of close to 3.4 million. Its capital city is Poznań; other important cities include Kalisz, Konin, Piła, Ostrów Wielkopolski and Gniezno (an early capital of Poland). It is bordered by seven other voivodeships: West Pomeranian to the northwest, Pomeranian to the north, Kuyavian-Pomeranian to the north-east, Łódź to the south-east, Opole to the south, Lower Silesian to the southwest and Lubusz to the west.

The city of Poznań has international twinning arrangements with the English county of Nottinghamshire.[3]

History

Greater Poland, sometimes called the "cradle of Poland," formed the heart of the 10th-century early Polish state. Poznań and Gniezno were early centers of royal power, but following the region's devastation by pagan rebellion in the 1030s, and an invasion by Bretislaus I of Bohemia in 1038, the capital was moved by Casimir the Restorer from Gniezno to Kraków.

Kórnik castle

In the testament of Bolesław III Krzywousty, which initiated the period of fragmentation of Poland (1138–1320), the western part of Greater Poland (including Poznań) was granted to Mieszko III the Old. The eastern part, with Gniezno and Kalisz, was part of the Duchy of Kraków, granted to Władysław II. However for most of the period the two parts were under a single ruler, and were known as the Duchy of Greater Poland (although at times there were separately ruled duchies of Poznań, Gniezno, Kalisz and Ujście). The region came under the control of Władysław I the Elbow-High in 1314, and thus became part of the reunited Poland of which Władyslaw was crowned king in 1320.

In the reunited kingdom, and later in the Polish–Lithuanian Commonwealth, the country came to be divided into administrative units called voivodeships. In the case of the Greater Poland region these were Poznań Voivodeship and Kalisz Voivodeship. The Commonwealth also had larger subdivisions known as *prowincja*, one of which was named Greater Poland. However, this *prowincja* covered a larger area than the Greater Poland region itself, also taking in Masovia and Royal Prussia. (This division of Crown Poland into two entities called Greater and Lesser Poland had its roots in the Statutes of Casimir the Great of 1346–1362, where the laws of "Greater Poland" – the northern part of the country – were codified in the Piotrków statute, with those of "Lesser Poland" in the separate Wiślica statute.)

In 1768 a new Gniezno Voivodeship was formed out of the northern part of Kalisz Voivodeship. However more far-reaching changes would come with the Partitions of Poland. In the first partition (1772), northern parts of Greater Poland along the Noteć (German *Netze*) were taken over by Prussia, becoming the Netze District. In the second partition (1793) the whole of Greater Poland was absorbed by Prussia, becoming part of the province of South Prussia. It remained so in spite of the first Greater Poland Uprising (1794), part of the unsuccessful Kościuszko Uprising directed chiefly against Russia.

More successful was the Greater Poland Uprising of 1806, which led to the region's becoming part of the Napoleonic Duchy of Warsaw (forming the Poznań Department and parts of the Kalisz and Bydgoszcz Departments). However, following the Congress of Vienna in 1815, Greater Poland was again partitioned, with the western part (including Poznań) going to Prussia. The eastern part joined the Russian-controlled Kingdom of Poland, where it formed the Kalisz Voivodeship until 1837, then the Kalisz Governorate (merged into the Warsaw Governorate between 1844 and 1867).

Soldiers in the Greater Poland Uprising of 1918–1919

Within the Prussian empire, western Greater Poland became the Grand Duchy of Posen (Poznań), which theoretically held some autonomy. Following an unrealized uprising in 1846, and the more substantial but still unsuccessful uprising of 1848 (during the Spring of Nations), the Grand Duchy was replaced by the Province of Posen. The authorities made efforts to Germanize the region, particularly after the founding of Germany in 1871, and from 1886 onwards the Prussian Settlement Commission was active in increasing German land ownership in formerly Polish areas.

Following the end of World War I, the Greater Poland Uprising (1918–1919) ensured that most of the region became part of the newly independent Polish state, forming most of Poznań Voivodeship (1921–1939). Northern and some western parts of Greater Poland remained in Germany, where they formed much of the province of Posen-West Prussia (1922–1938), whose capital was Schneidemühl (Piła).

Following the German invasion of 1939, Greater Poland was incorporated into Nazi Germany, becoming the province called Reichsgau Posen, later Reichsgau Wartheland (*Warthe* being the German name for the Warta river). The Polish population was oppressed, with many former officials and others considered potential enemies by the Nazis being imprisoned or executed, including at the notorious Fort VII concentration camp in Poznań. Poznań was declared a stronghold city *(Festung)* in the closing stages of the war, being taken by the Red Army in the Battle of Poznań, which ended on 22 February 1945.

After the war, Greater Poland was fully within the Polish People's Republic, as Poznań Voivodeship. With the reforms of 1975 this was divided into smaller provinces (the voivodeships of Kalisz, Konin, Leszno and Piła, and a smaller Poznań Voivodeship). The present-day Greater Poland Voivodeship, again with Poznań as its capital, was created in 1999.

Cities and towns

The voivodeship contains 109 cities and towns. These are listed below in descending order of population (according to official figures for 2006[4]):

Poznań is the capital of the Greater Poland Voivodeship

1. Poznań (566,546)
2. Kalisz (108,575)
3. Konin (80,618)
4. Piła (75,044)
5. Ostrów Wielkopolski (72,577)
6. Gniezno (70,080)
7. Leszno (64,079)
8. Śrem (30,227)
9. Swarzędz (29,894)
10. Krotoszyn (29,421)
11. Turek (29,302)
12. Września (28,617)
13. Luboń (26,935)
14. Jarocin (25,834)
15. Wągrowiec (24,681)
16. Kościan (24,102)
17. Koło (23,034)

18. Środa Wielkopolska (21,635)
19. Rawicz (21,301)
20. Gostyń (20,588)
21. Chodzież (19,652)
22. Szamotuły (18,760)
23. Złotów (18,468)
24. Oborniki (17,850)
25. Pleszew (17,787)
26. Trzcianka (16,756)
27. Nowy Tomyśl (15,225)
28. Kępno (14,710)
29. Ostrzeszów (14,536)
30. Słupca (14,363)
31. Grodzisk Wielkopolski (13,703)
32. Wolsztyn (13,557)
33. Mosina (12,150)
34. Wronki (11,551)
35. Czarnków (11,356)
36. Międzychód (10,920)
37. Rogoźno (10,905)

Kalisz Old Town Hall

Gniezno with its cathedral is the seat of the catholic Primate of Poland

1. Murowana Goślina (10,140)
2. Puszczykowo (9,311)
3. Opalenica (9,104)
4. Kostrzyn (8,539)
5. Jastrowie (8,403)
6. Pobiedziska (8,329)
7. Ujście (8,134)
8. Witkowo (7,855)
9. Trzemeszno (7,789)
10. Pniewy (7,464)
11. Zbąszyń (7,300)
12. Kórnik (6,981)
13. Kłodawa (6,829)
14. Koźmin Wielkopolski (6,707)
15. Krzyż Wielkopolski (6,283)
16. Buk (6,181)
17. Sieraków (5,994)
18. Wieleń (5,940)
19. Śmigiel (5,452)
20. Stęszew (5,339)
21. Wyrzysk (5,234)
22. Czempiń (5,135)
23. Nowe Skalmierzyce (5,080)

24. Odolanów (4,960)
25. Zduny (4,498)
26. Golina (4,330)
27. Szamocin (4,267)
28. Kleczew (4,173)
29. Krobia (4,022)
30. Skoki (3,866)
31. Sompolno (3,695)
32. Krajenka (3,651)
33. Miłosław (3,589)
34. Tuliszków (3,393)
35. Okonek (3,357)
36. Gołańcz (3,342)
37. Rakoniewice (3,253)

1. Nekla (3,203)
2. Pyzdry (3,188)
3. Łobżenica (3,172)
4. Miejska Górka (3,128)
5. Ślesin (3,102)
6. Kobylin (3,084)
7. Bojanowo (3,014)
8. Margonin (2,956)
9. Zagórów (2,932)
10. Lwówek (2,909)
11. Poniec (2,875)
12. Sulmierzyce (2,772)
13. Wysoka (2,750)
14. Książ Wielkopolski (2,724)
15. Kłecko (2,677)
16. Czerniejewo (2,556)
17. Rydzyna (2,539)
18. Borek Wielkopolski (2,486)
19. Rychwał (2,377)
20. Obrzycko (2,170)
21. Dąbie (2,087)
22. Żerków (2,058)
23. Raszków (2,037)
24. Osieczna (2,018)
25. Ostroróg (1,995)
26. Pogorzela (1,974)
27. Grabów nad Prosną (1,967)
28. Jutrosin (1,872)
29. Mikstat (1,840)
30. Przedecz (1,771)
31. Wielichowo (1,765)
32. Stawiszyn (1,554)
33. Krzywiń (1,547)

34. Dobra (1,511)
35. Dolsk (1,479)

.

Geography

Topography

The relief of Greater Poland, geological conditions and soil have been shaped by two glaciations:

- The Baltic glaciation in the lowlands of northern and central Europe where there are now numerous lakes of the Pomeranian Lake District, a feature especially common in and around Poznań and Gniezno.
- The Mid-glaciation in the southern part of the province, where there is less terrain diversity and a lack of major lakes.

Lake Góreckie in Wielkopolska National Park

The highest elevation is Greater Kobyla Mountain (284 m) in the Ostrzeszowski Hills, the lowest area is located in the valley of the Warta River at the mouth of its tributary the Noteć (21 m) in the north-western part of the region. Agriculturally fertile soils account for around 60% of the province's area, while 20%, the rest of the non-forested or urban areas, is mostly wetland soil (muck-peat and alluvial soils).

An area of approximately 800 thousand hectares is covered by forests, this represents around 25.8% of the total surface area of the region. In the lake districts of the northern and central parts of the province there are about 800 lakes; 58% of which cover an area of at least 10 hectares and 8%, with an area exceeding 100 hectares. The largest reservoir is the natural Greater Powidzkie Lake (1036 ha) in the Gniezno Lake District.

Wielkopolska Region lies within the basin of the Oder River, 88% of the province's surface water drains into the Warta river basin, and the remaining 12% is drained by a multitude of other river systems, including the Barycz, Ladislaus Trench and Obrzycy waterways. The quality of river waters is generally poor, but their condition is gradually improving and should soon be classed as 'clean'.

Geology

A modern coal-fired power plant in Pątnów

The main mineral energy resources in Greater Poland are lignite, natural gas, oil and peat.

Brown coal deposits are currently mined in the Konin area, and form the basis for the province's power industry (the Pątnów-Adams-Konin coal-fired power stations account for more than 10% of the national electricity production). The region also has significant quantities of peat deposits; it is calculated that there are ca. 886 thousand hectares of land covered with an average thickness of 1.5 m of peat. An abundance of raw materials used in the production of numerous medicines was recently discovered in the muds of Błażejewo, Oderbank and Mechnacz. In addition, very large deposits of brown coal have been discovered in the vicinity of Kościan, these however are not currently being extracted and probably never will be extracted, due to the expense that would be incurred in adapting the site to build a coal mine and the need to resettle thousands of people.

Rock salt is mined intensively at a salt mine in Kłodawa (this mine alone accounts for about 20% of domestic production).

Throughout the province there are significant deposits of aggregates, gypsum, ceramic materials, and lacustrine chalk. In Kościan the largest and most modern, a natural gas production site is in operation. It supplies raw material for Kościańska Zieme, and Zielona Gora CHP. It is estimated that at the rate local gas reserves are being exploited, the reserves in Kościan will be enough for about 20 years of operation, thus practically allowing for local independence against the effects of gas crises.

Climate

Wielkopolska is influenced by oceanic air masses that affect the mildness of the climate. The farther east one travels the more distinctly continental the climate becomes. The area is situated in the Silesian Greater Poland agro-climatic region where the average annual temperature is about 8.2 °C, and in the north drops to around 7.6 °C. It is slightly warmer in the south and west where the average temperature is usually about 8.5 °C. The number of days with snow can reach up to 57 days in and around the Kalisz district.

The growing season is one of the longest in Poland. On the province's southern plains this season constitutes around 228 days, while north of Gniezno and Szamotuł this gradually declines to 216 days.

Precipitation ranges from 500 to 550 mm. Despite this the region is still faced with a deficit in rainfall, particularly in the eastern part of the province (around Słupcy, Kazimierz Biskupi, Kleczew) where sometimes experience only 450 mm of rainfall per year, this threatens steppization of the region. Throughout the province there is typically a prevailing westerly wind.

Transportation

The A2 motorway traverses the voivodeship

Greater Poland is a major transport hub within Poland, a great deal of traffic from Russia and other states of the former Soviet Union pass through Poznań and Konin to reach Germany and other EU member states. To the south runs the international route from Gdańsk via Poznań and Leszno to Prague and then to the south of Europe. There is also a major highway being built in in the province, the A2 motorway, which when completed will run from the western border of Poland with Germany, through Poznań to Warsaw and then onwards via Belarus to Moscow.

The main railway hubs located in Greater Poland are Poznań, Piła and Ostrów Wielkopolski. PKP Intercity operate a number of trains a day between Warsaw and Berlin which provide a fast connection for the two cities also to Poznań. This route was the first in Poland, adapted for use by the European high-speed transportation system. In the near future the government expects to construct a high-speed rail line in the shape of a Y connecting Kalisz and Poznań from Łódź, Warsaw and Wrocław.

Poznań is the port of arrival for most international travellers as it is plays host to Ławica International Airport which has recently seen the second highest passenger growth rate in the country.

Politics

The Greater Poland voivodeship's government is headed by the province's voivode *(governor)* who is appointed by the Polish Prime Minister. The voivode is then assisted in performing his duties by the voivodeship's marshal, who is the appointed speaker for the voivodeship's executive and is elected by the sejmik *(provincial assembly)*. The current voivode of Greater Poland is Piotr Florek, whilst the present marshal is Marek Woźniak.

The Sejmik of Greater Poland consists of 39 members.

Current voivode of Greater Poland, Piotr Florek (front, 2nd from right), with marshal Marek Woźniak (front, 3rd from right)

Kuyavian-Pomeranian Regional Assembly elections on 21 November 2010[5]			
Party	**Votes**	**%**	**Total seats held**
Civic Platform (PO)	345,209	32.04	17
Law and Justice (PiS)	193,395	17.95	6
Left and Democrats (LiD)	232,704	21.60	9
Polish People's Party (PSL)	193,953	18.00	7
Others	112,171	10.41	0
Total	**1,077,356**	**100.00**	**39**

- Votes counted: 1,271,549
- Valid votes: 1,077,356
- **Turnout: 47.04%**

Administrative division

Greater Poland Voivodeship is divided into 35 counties (powiats): 4 city counties and 31 land counties. These are further divided into 226 gminas.

The counties are listed in the following table (ordering within categories is by decreasing population).

English and Polish names	Area		Population (2006)	Seat	Other towns	Total gminas
	(km²)	(sq mi)				
City counties						
Poznań	262	101	566,546			1
Kalisz	70	27	108,575			1
Konin	82	32	80,618			1
Leszno	32	12	64,079			1
Land counties						
Poznań County powiat poznański	1900	734	291,562	Poznań *	Swarzędz, Luboń, Mosina, Murowana Goślina, Puszczykowo, Kostrzyn, Pobiedziska, Kórnik, Buk, Stęszew	17
Ostrów Wielkopolski County powiat ostrowski	1161	448	158,407	Ostrów Wielkopolski	Nowe Skalmierzyce, Odolanów, Raszków	8
Gniezno County powiat gnieźnieński	1254	484	140,333	Gniezno	Witkowo, Trzemeszno, Kłecko, Czerniejewo	10
Piła County powiat pilski	1267	489	137,099	Piła	Wyrzysk, Ujście, Łobżenica, Wysoka	9
Konin County powiat koniński	1579	610	123,646	Konin *	Golina, Kleczew, Sompolno, Ślesin, Rychwał	14
Koło County powiat kolski	1011	390	88,601	Koło	Kłodawa, Dąbie, Przedecz	11
Czarnków-Trzcianka County powiat czarnkowsko-trzcianecki	1808	698	86,134	Czarnków	Trzcianka, Krzyż Wielkopolski, Wieleń	8
Szamotuły County powiat szamotulski	1120	432	85,849	Szamotuły	Wronki, Pniewy, Obrzycko, Ostroróg	8
Turek County powiat turecki	929	359	83,635	Turek	Tuliszków, Dobra	9
Kalisz County powiat kaliski	1160	448	80,369	Kalisz *	Stawiszyn	11
Kościan County powiat kościański	723	279	77,760	Kościan	Śmigiel, Czempiń, Krzywiń	5
Krotoszyn County powiat krotoszyński	714	276	77,092	Krotoszyn	Koźmin Wielkopolski, Zduny, Kobylin, Sulmierzyce	6
Gostyń County powiat gostyński	810	313	75,683	Gostyń	Krobia, Poniec, Borek Wielkopolski, Pogorzela	7
Września County powiat wrzesiński	704	272	73,778	Września	Miłosław, Nekla, Pyzdry	5

Nowy Tomyśl County powiat nowotomyski	1012	391	71,817	Nowy Tomyśl	Opalenica, Zbąszyń, Lwówek	6
Jarocin County powiat jarociński	588	227	70,390	Jarocin	Żerków	4
Złotów County powiat złotowski	1661	641	68,526	Złotów	Jastrowie, Okonek, Krajenka	8
Wągrowiec County powiat wągrowiecki	1041	402	67,606	Wągrowiec	Skoki, Gołańcz	7
Pleszew County powiat pleszewski	712	275	61,951	Pleszew		6
Rawicz County powiat rawicki	553	214	59,375	Rawicz	Miejska Górka, Bojanowo, Jutrosin	5
Słupca County powiat słupecki	838	324	58,725	Słupca	Zagórów	8
Śrem County powiat śremski	574	222	58,646	Śrem	Książ Wielkopolski, Dolsk	4
Oborniki County powiat obornicki	713	275	55,976	Oborniki	Rogoźno	3
Kępno County powiat kępiński	608	235	55,335	Kępno		7
Wolsztyn County powiat wolsztyński	680	263	54,718	Wolsztyn		3
Środa Wielkopolska County powiat średzki	623	241	54,568	Środa Wielkopolska		5
Ostrzeszów County powiat ostrzeszowski	772	298	54,490	Ostrzeszów	Grabów nad Prosną, Mikstat	7
Leszno County powiat leszczyński	805	311	50,024	Leszno *	Rydzyna, Osieczna	7
Grodzisk Wielkopolski County powiat grodziski	644	249	49,444	Grodzisk Wielkopolski	Rakoniewice, Wielichowo	5
Chodzież County powiat chodzieski	681	263	46,967	Chodzież	Szamocin, Margonin	5
Międzychód County powiat międzychodzki	737	285	36,329	Międzychód	Sieraków	4
* seat not part of the county						

Protected areas

Protected areas in Greater Poland Voivodeship include two National Parks and 12 Landscape Parks. These are listed below.

- Drawno National Park (partly in Lubusz and West Pomeranian Voivodeships)
- Wielkopolska National Park
- Barycz Valley Landscape Park (partly in Lower Silesian Voivodeship)
- Chłapowski Landscape Park
- Lednica Landscape Park
- Powidz Landscape Park
- Promno Landscape Park
- Przemęt Landscape Park (partly in Lubusz Voivodeship)
- Pszczew Landscape Park (partly in Lubusz Voivodeship)
- Puszcza Zielonka Landscape Park
- Rogalin Landscape Park
- Sieraków Landscape Park
- Warta Landscape Park
- Żerków-Czeszewo Landscape Park

Most common surnames in the region

1. Nowak: 35,011
2. Kaczmarek: 24,185
3. Wojciechowski: 12,928

See also

- Prussia's Province of Posen (1818–1919)
- Second Polish Republic's Poznań Voivodeship (1921–1939)

References

[1] http://www.en.poznan.uw.gov.pl/
[2] Arkadiusz Belczyk, Tłumaczenie polskich nazw geograficznych na język angielski (http://serwistlumacza.com/content/view/27/32/) [Translation of Polish Geographical Names into English], 2002-2006.
[3] http://www.nottinghamshire.gov.uk/home/your_council/howweprovideyourservices/partnerships/internationalandtwinning/transnationalpartnerships.htm
[4] http://www.stat.gov.pl/gus/45_655_PLK_HTML.htm
[5] "Greater Poland Regional Assembly elections" (http://wybory2010.pkw.gov.pl/geo/pl/300000/300000.html#tabs-5). State Electoral Commission. . Retrieved 2011-05-28.

Further reading

- Zygmunt Boras, *Książęta Piastowscy Wielkopolski* (Piast Princes of Wielkopolska), Poznań, Wydawnictwo Poznańskie, 1983.

External links

- wielkopolska-region.pl (http://www.wielkopolska-region.pl/) Information about Great Poland; Multilingual.
- greatpoland.eu (http://www.greatpoland.eu/) Data base about: Business, Culture, Sport, Motorization, Tourism, Agrotourism, Medicine, Health by CyberWielkopolska
- Greater Poland Local Government Office (http://www.wielkopolska.mw.gov.pl/)
- ChefMoz Dining Guide Greater Poland (http://chefmoz.org/Poland/WP/)
- Wielkopolska (http://www.dmoz.org/Regional/Europe/Poland/Voivodships/Wielkopolska/) at the Open Directory Project

Voivodeships_of_Poland

The **voivodeship,**[1] or **province,**[2] called in Polish ***województwo*** [vɔjɛˈvut͡stfɔ] (plural *województwa*), has been a high-level administrative subdivision of Poland since the 14th century.

The Polish local government reforms adopted in 1998, which went into effect on 1 January 1999, created sixteen new voivodeships. These replaced the 49 former voivodeships that had existed from 1 July 1975.

Today's voivodeships are mostly named after historical and geographical regions, while those prior to 1998 generally took their names from the cities on which they were centered. The new units range in area from under 10000 km^2 (**unknown operator: u'strong'** sq mi) (Opole Voivodeship) to over 35000 km^2 (**unknown operator: u'strong'** sq mi) (Masovian Voivodeship), and in population from one million (Lubusz Voivodeship) to over five million (Masovian Voivodeship).

Administrative authority at voivodeship level is shared between a government-appointed governor called a voivode (Polish *wojewoda*), an elected assembly called a sejmik, and an executive chosen by that assembly. The leader of that executive is called the *marszałek województwa* (voivodeship marshal). Voivodeships are further divided into powiats (counties) and gminas (communes or municipalities): see Administrative division of Poland.

Voivodeships since 1999

Administrative powers

Competences and powers at voivodeship level are shared between the voivode (governor), the sejmik (regional assembly) and the executive. In most cases these institutions are all based in one city, but in Kuyavian-Pomeranian and Lubusz Voivodeship the voivode's offices are in a different city from those of the executive and the sejmik. Voivodeship capitals are listed in the table below.

The **voivode** is appointed by the Prime Minister and is the regional representative of the central government. The voivode acts as the head of central government institutions at regional level (such as the police and fire services, passport offices, and various inspectorates), manages central government property in the region, oversees the functioning of local government, coordinates actions in the field of public safety and environment protection, and exercises special powers in emergencies. The voivode's offices collectively are known as the *urząd wojewódzki.*

The **sejmik** is elected every four years, at the same time as the local authorities at powiat and gmina level. It passes bylaws, including the voivodeship's development strategies and budget. It also elects the *marszałek* and other members of the executive, and holds them to account.

The **executive** (*zarząd województwa*), headed by the *marszałek*, drafts the budget and development strategies, implements the resolutions of the sejmik, manages the voivodeship's property, and deals with many aspects of regional policy, including management of European Union funding. Its offices collectively are known as the *urząd marszałkowski*.

Map and table of voivodeships

Polish voivodeships since 1999

Abbreviation	Coat of arms	code	car plates	Voivodeship	Capital	Area km^2	Population (December 31, 2003)	Population (June 30, 2004)
DS		02	D	Lower Silesian (*dolnośląskie*)	Wrocław	19 947.76	2 898 313	2 895 729
KP		04	C	Kuyavian-Pomeranian (*kujawsko-pomorskie*)	Bydgoszcz[1] Toruń[2]	17 969.72	2 068 142	2 067 548
LU		06	L	Lublin (*lubelskie*)	Lublin	25 114.48	2 191 172	2 187 918
LB		08	F	Lubusz (*lubuskie*)	Gorzów Wielkopolski[1] Zielona Góra[2]	13 984.44	1 008 786	1 009 177
LD		10	E	Łódź (*łódzkie*)	Łódź	18 219.11	2 597 094	2 592 568
MP		12	K	Lesser Poland (*małopolskie*)	Kraków	15 144.10	3 252 949	3 256 171
MA		14	W	Masovian (*mazowieckie*)	Warsaw	35 597.80	5 135 732	5 139 545
OP		16	O	Opolskie	Opole	9 412.47	1 055 667	1 053 723
PK		18	R	Podkarpackie	Rzeszów	17 926.28	2 097 248	2 097 325
PD		20	B	Podlaskie (*podlaskie*)	Białystok	20 179.58	1 205 117	1 204 036
PM		22	G	Pomeranian (*pomorskie*)	Gdańsk	18 292.88	2 188 918	2 192 404

SL		24	S	Silesian (*śląskie*)	Katowice	12 294.04	4 714 982	4 707 825
SW		26	T	Świętokrzyskie	Kielce	11 672.34	1 291 598	1 290 176
WM		28	N	Warmian-Masurian (*warmińsko-mazurskie*)	Olsztyn	24 202.95	1 428 885	1 428 385
WP		30	P	Greater Poland (*wielkopolskie*)	Poznań	29 825.59	3 359 932	3 362 011
ZP		32	Z	West Pomeranian (*zachodniopomorskie*)	Szczecin	22 901.48	1 696 073	1 695 708
(¹) – seat of voivode, (²) – seat of sejmik and marszałek								

See also:

- *Map of Polish Regions* [3]
- *Administrative division of Poland (from Commission on Standardization of Geographical Names Outside Poland website, in English)* [4]
- *Official map by Head Office of Geodesy and Cartography* [5]

Former voivodeships

Poland's voivodeships 1975–1998

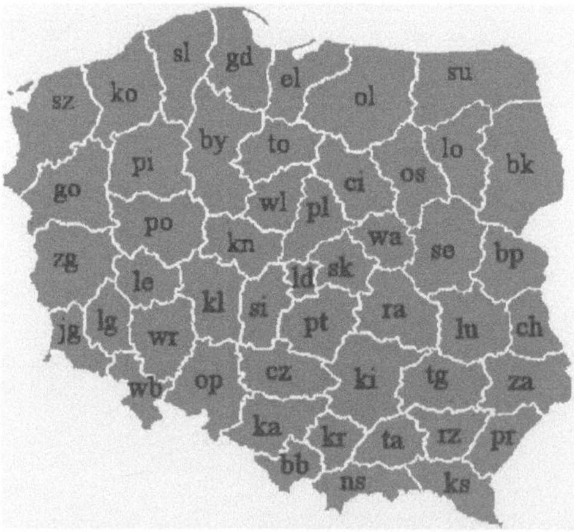

Administrative division of Poland between 1979 and 1998 included 49 voivodeships upheld after the establishment of the Third Polish Republic in 1989 for another decade. This reorganization of administrative division of Poland was mainly a result of local government reform acts of 1973–1975. In place of the three-level administrative division (voivodeship, county, commune), a new two-level administrative division was introduced (49 small voivodeships, and communes). The three smallest voivodeships – Warsaw, Kraków and Łódź – had the special status of municipal voivodeship; the city president (mayor) was also provincial governor.

Polish voivodeships and separate cities 1975-1998							
Abbreviation	Voivodeship	Polish name	Capital	Area km² (1998)	Population (1980)	No. of cities	No. of communes
bp	Biała Podlaska Voivodeship	bialskopodlaskie	Biała Podlaska	5 348	286 400	6	35
bk	Białystok Voivodeship	białostockie	Białystok	10 055	641 100	17	49
bb	Bielsko-Biała Voivodeship	bielskie	Bielsko-Biała	3 704	829 900	18	47
by	Bydgoszcz Voivodeship	bydgoskie	Bydgoszcz	10 349	1 036 000	27	55
ch	Chełm Voivodeship	chełmskie	Chełm	3 865	230 900	4	25
ci	Ciechanów Voivodeship	ciechanowskie	Ciechanów	6 362	405 400	9	45
cz	Częstochowa Voivodeship	częstochowskie	Częstochowa	6 182	747 900	17	49
el	Elbląg Voivodeship	elbląskie	Elbląg	6 103	441 500	15	37
gd	Gdańsk Voivodeship	gdańskie	Gdańsk	7 394	1 333 800	19	43
go	Gorzów Voivodeship	gorzowskie	Gorzów Wielkopolski	8 484	455 400	21	38
jg	Jelenia Góra Voivodeship	jeleniogórskie	Jelenia Góra	4 378	492 600	24	28
kl	Kalisz Voivodeship	kaliskie	Kalisz	6 512	668 000	20	53
ka	Katowice Voivodeship	katowickie	Katowice	6 650	3 733 900	43	46
ki	Kielce Voivodeship	kieleckie	Kielce	9 211	1 068 700	17	69
kn	Konin Voivodeship	konińskie	Konin	5 139	441 200	18	43
ko	Koszalin Voivodeship	koszalińskie	Koszalin	8 470	462 200	17	35
kr	Kraków Voivodeship	krakowskie	Kraków	3 254	1 167 500	10	38
ks	Krosno Voivodeship	krośnieńskie	Krosno	5 702	448 200	12	37
lg	Legnica Voivodeship	legnickie	Legnica	4 037	458 900	11	31
le	Leszno Voivodeship	leszczyńskie	Leszno	4 254	357 600	19	28
lu	Lublin Voivodeship	lubelskie	Lublin	6 793	935 200	16	62
lo	Łomża Voivodeship	łomżyńskie	Łomża	6 684	325 800	12	39
ld	Łódź Voivodeship	łódzkie	Łódź	1523	1 127 800	8	11
ns	Nowy Sącz Voivodeship	nowosądeckie	Nowy Sącz	5 576	628 800	14	41
ol	Olsztyn Voivodeship	olsztyńskie	Olsztyn	12 327	681 400	21	48
op	Opole Voivodeship	opolskie	Opole	8 535	975 000	29	61
os	Ostrołęka Voivodeship	ostrołęckie	Ostrołęka	6 498	371 400	9	38
pi	Piła Voivodeship	pilskie	Piła	8 205	437 100	24	35
pt	Piotrków Voivodeship	piotrkowskie	Piotrków Trybunalski	6 266	604 200	10	51
pl	Płock Voivodeship	płockie	Płock	5 117	496 100	9	44
po	Poznań Voivodeship	poznańskie	Poznań	8 151	1 237 800	33	57
pr	Przemyśl Voivodeship	przemyskie	Przemyśl	4 437	380 000	9	35
ra	Radom Voivodeship	radomskie	Radom	7 295	702 300	15	61
rz	Rzeszów Voivodeship	rzeszowskie	Rzeszów	4 397	648 900	13	41

se	Siedlce Voivodeship	siedleckie	Siedlce	8 499	616 300	12	66
si	Sieradz Voivodeship	sieradzkie	Sieradz	4 869	392 300	9	40
sk	Skierniewice Voivodeship	skierniewickie	Skierniewice	3 959	396 900	8	36
sl	Słupsk Voivodeship	słupskie	Słupsk	7 453	369 800	11	31
su	Suwałki Voivodeship	suwalskie	Suwałki	10 490	422 600	14	42
sz	Szczecin Voivodeship	szczecińskie	Szczecin	9 981	897 900	29	50
tg	Tarnobrzeg Voivodeship	tarnobrzeskie	Tarnobrzeg	6 283	556 300	14	46
ta	Tarnów Voivodeship	tarnowskie	Tarnów	4 151	607 000	9	41
to	Toruń Voivodeship	toruńskie	Toruń	5 348	610 800	13	41
wb	Wałbrzych Voivodeship	wałbrzyskie	Wałbrzych	4 168	716 100	31	30
wa	Warsaw Voivodeship	warszawskie	Warsaw	3 788	2 319 100	27	32
wl	Włocławek Voivodeship	włocławskie	Włocławek	4 402	413 400	14	30
wr	Wrocław Voivodeship	wrocławskie	Wrocław	6 287	1 076 200	16	33
za	Zamość Voivodeship	zamojskie	Zamość	6 980	472 100	5	47
zg	Zielona Góra Voivodeship	zielonogórskie	Zielona Góra	8 868	609 200	26	50

Poland's voivodeships 1945–1975

After World War II, the new administrative division of the country within the new national borders was based on the prewar one and included 14 (+2) voivodeships, then 17 (+5). The voivodeships in the east that had not been annexed by the Soviet Union had their borders left almost unchanged. The newly acquired territories in the west and north were organized into the new voivodeships of Szczecin, Wrocław and Olsztyn, and partly joined to Gdańsk, Katowice and Poznań voivodeships. Two cities were granted voivodeship status: Warsaw and Łódź.

In 1950, new voivodeships were created: Koszalin (previously part of Szczecin), Opole (previously part of Katowice), and Zielona Góra (previously part of Poznań, Wrocław and Szczecin voivodeships). In addition, three more cities were granted voivodeship status: Wrocław, Kraków and Poznań.

Polish administrative division 1945-1975				
Car plates (since 1956)	**Voivodeship**	**Capital**	**Area km² (1965)**	**Population (1965)**
A	białostockie	Białystok	23 136	1 160 400
B	bydgoskie	Bydgoszcz	20 794	1 837 100
G	gdańskie	Gdańsk	10 984	1 352 800
S	katowickie	Katowice	9 518	3 524 300
C	kieleckie	Kielce	19 498	1 899 100
E	koszalińskie [1]	Koszalin	17 974	755 100
K	krakowskie	Kraków	15 350	2 127 600
F	Łódzkie	Łódź	17 064	1 665 200
L	lubelskie	Lublin	24 829	1 900 500
O	olsztyńskie	Olsztyn	20 994	956 600
H	opolskie [1]	Opole	9 506	1 009 200
P	poznańskie	Poznań	26 723	2 126 300
R	rzeszowskie	Rzeszów	18 658	1 692 800
M	szczecińskie	Szczecin	12 677	847 600
T	warszawskie	Warsaw	29 369	2 453 000
X	wrocławskie	Wrocław	18 827	1 967 000
Z	zielonogórskie [1]	Zielona Góra	14 514	847 200
car plates (since 1956)	**Separate city**		**Area km² (1965)**	**Population (1965)**
I	Łódź		214	744 100
W	Warsaw		446	1 252 600
?	Kraków [2]		230	520 100
?	Poznań [2]		220	438 200
?	Wrocław [2]		225	474 200
[1] – new voivodeships created in 1950; [2] – cities separated in 1957				

Poland's voivodeships 1921–1939

The administrative division of Poland in the interwar period included 16 voivodeships and Warsaw (with voivodeship rights).

Polish voivodeships in the interbellum (data as per April 1, 1937)					
car plates (since 1937)	Voivodeship Separate city	Capital	Area in 1000 km² (1930)	Population in 1000 (1931)	Map
00-19	City of Warsaw	Warsaw	0.14	1179.5	
85-89	warszawskie	Warsaw	31.7	2460.9	
20-24	białostockie	Białystok	26.0	1263.3	
25-29	kieleckie	Kielce	22.2	2671.0	
30-34	krakowskie	Kraków	17.6	2300.1	
35-39	lubelskie	Lublin	26.6	2116.2	
40-44	lwowskie	Lwów	28.4	3126.3	
45-49	łódzkie	Łódź	20.4	2650.1	
50-54	nowogródzkie	Nowogródek	23.0	1057.2	
55-59	poleskie	Brześć nad Bugiem	36.7	1132.2	
60-64	pomorskie	Toruń	25.7	1884.4	
65-69	poznańskie	Poznań	28.1	2339.6	
70-74	stanisławowskie	Stanisławów	16.9	1480.3	
75-79 ?	śląskie	Katowice	5.1	1533.5	
80-84	tarnopolskie	Tarnopol	16.5	1600.4	
90-94	wileńskie	Wilno	29.0	1276.0	
95-99	wołyńskie	Łuck	35.7	2085.6	

Congress Poland 1816–1837

From 1816 to 1837 there were 8 voivodeships in Congress Poland.

- Augustów Voivodeship
- Kalisz Voivodeship
- Kraków Voivodeship
- Lublin Voivodeship
- Mazowsze Voivodeship
- Płock Voivodeship
- Podlasie Voivodeship
- Sandomierz Voivodeship

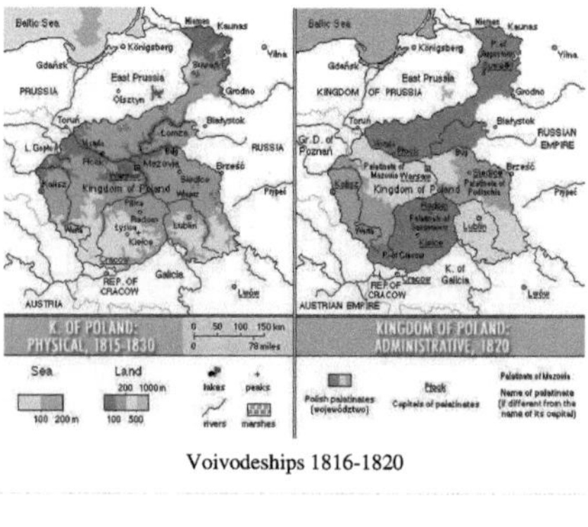

Voivodeships 1816-1820

Polish–Lithuanian Commonwealth 1569–1795

Greater Poland (*Wielkopolska*)

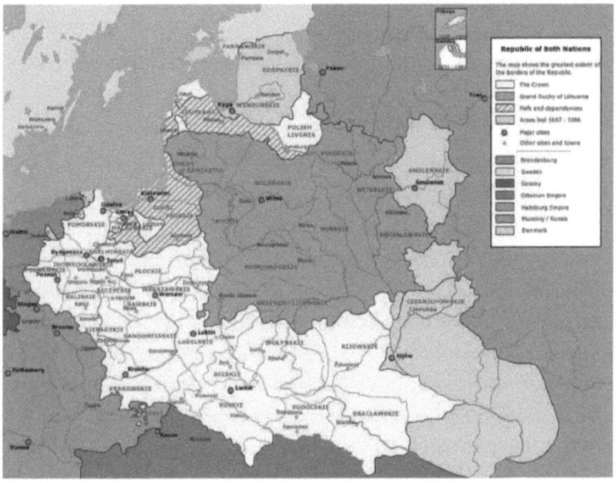

Voivodeships of the Republic of the Two Nations ("Polish–Lithuanian Commonwealth").

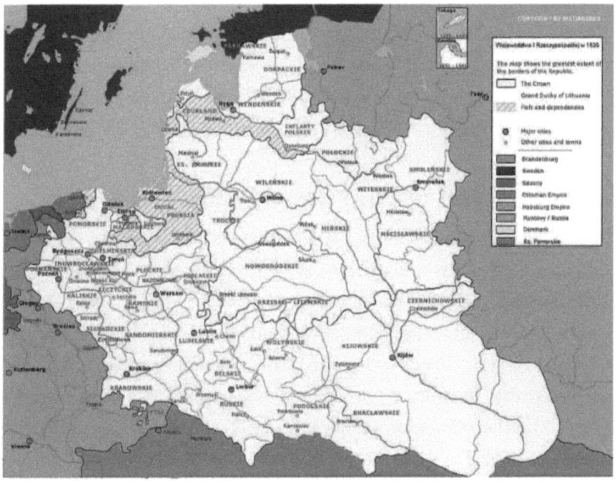

Voivodeships of the Commonwealth of the Two Nations in 1635

- Poznań Voivodeship (*województwo poznańskie*, Poznań)
- Kalisz Voivodeship (*województwo kaliskie*, Kalisz)
- Gniezno Voivodeship (*województwo gnieźnieńskie*, Gniezno) from 1768
- Sieradz Voivodeship (*województwo sieradzkie*, Sieradz)
- Łęczyca Voivodeship (*województwo łęczyckie*, Łęczyca)
- Brześć Kujawski Voivodeship (*województwo brzesko-kujawskie*, Brześć Kujawski)
- Inowrocław Voivodeship (*województwo inowrocławskie*, Inowrocław)
- Chełmno Voivodeship (*województwo chełmińskie*, Chełmno)
- Malbork Voivodeship (*województwo malborskie*, Malbork)
- Pomeranian Voivodeship (*województwo pomorskie*, Gdańsk)
- Duchy of Warmia (*Księstwo Warmińskie*, Lidzbark Warmiński)
- Duchy of Prussia (*Księstwo Pruskie*, Królewiec) (Then Königsberg)
- Płock Voivodeship (*województwo płockie*, Płock)
- Rawa Voivodeship (*województwo rawskie*, Rawa Mazowiecka)
- Masovian Voivodeship (*województwo mazowieckie*, Warszawa)

Lesser Poland (*Małopolska*)

- Kraków Voivodeship (*województwo krakowskie*, Kraków)
- Sandomierz Voivodeship (*województwo sandomierskie*, Sandomierz)
- Lublin Voivodeship (*województwo lubelskie*, Lublin)
- Podlaskie Voivodeship (*województwo podlaskie*, Drohiczyn)
- Ruthenian Voivodeship (*województwo ruskie*, Lwów)
- Bełz Voivodeship (*województwo bełzkie*, Bełz)

- Volhynian Voivodeship (*województwo wołyńskie*, Łuck)
- Podole Voivodeship (*województwo podolskie*, Kamieniec Podolski)
- Bracław Voivodeship (*województwo bracławskie*, Bracław)
- Kijów Voivodeship (*województwo kijowskie*, Kijów)
- Czernihów Voivodeship (*województwo czernichowskie*, Czernihów)

Grand Duchy of Lithuania

Here the first name given is English, then in brackets – Lithuanian, and then Polish.

- Wilno Voivodeship (*Vilniaus vaivadija, województwo wileńskie*, Vilnius)
- Troki Voivodeship (*Trakų vaivadija, województwo trockie*, Trakai)
- Nowogródek Voivodeship (*Naugarduko vaivadija, województwo nowogrodzkie*, Nowogródek)
- Brest-Litovsk Voivodeship (*Lietuvos Brastos vaivadija, województwo brzesko-litewskie*, Brześć Litewski)
- Minsk Voivodeship (*Minsko vaivadija, województwo mińskie*, Mińsk)
- Mścisław Voivodeship (*Mstslavlio vaivadija, województwo mścisławskie*, Mscislaw)
- Smolensk Voivodeship (*Smolensko vaivadija, województwo smoleńskie*, Smoleńsk)
- Vitebsk Voivodeship (*Vitebsko vaivadija, województwo witebskie*, Witebsk)
- Połock Voivodeship (*Polocko vaivadija, województwo połockie*, Połock)
- Duchy of Samogita (*Žemaičių seniūnija, księstwo żmudzkie*, Varniai then called Medininkai)

Duchy of Livonia

- Wenden Voivodeship (*województwo wendeńskie*, Wenden) since 1598 till the 1620s
- Dorpat Voivodeship (*województwo dorpackie*, Dorpat) since 1598 till the 1620s
- Parnawa Voivodeship (*województwo parnawskie*, Parnawa) since 1598 till the 1620s
- Inflanty Voivodeship (*województwo inflanckie*, Dyneburg) since the 1620s
- Duchy of Courland and Semigalia (*księstwo Kurlandii i Semigalii*, Mitawa)

Etymology and use of "voivodeship"

Some English-language sources, in historic contexts, speak of "palatinates" rather than "voivodeships"; the former term traces back to the Latin *palatinus* ("palatine"). More commonly used now is "voivodeship", a loanword-calque hybrid formed on the Polish "*województwo*". Other sources refer instead to "provinces" (Polish singular: "*prowincja*"), though in pre-1795 contexts this may be confusing because the cognate Polish "*prowincyja*" (as it was then spelled) was idiosyncratically applied, until the last of the three Partitions of the Polish–Lithuanian Commonwealth, in 1795, to each of the three main Regions (Greater Poland, Lesser Poland, and Lithuania) of the Polish–Lithuanian Commonwealth, each of those Regions in turn comprising a number of *województwa* (plural of "*województwo*").

The Polish "*województwo*", designating a second-tier Polish or Polish–Lithuanian administrative unit, derives from "*wojewoda*" (etymologically, a "war leader" or "leader of warriors", but now simply the governor of a *województwo*) and the suffix "*-stwo*" (a "state or condition").

The English "voivodeship", which is a hybrid of the loanword "voivode" and "-ship" (the latter a suffix, likewise meaning a "state or condition", that calques the Polish "*-stwo*"), has never been much used and is absent from many dictionaries. According to the *Oxford English Dictionary*, it first appeared in 1792, spelled "woiwodship", in the sense of "the district or province governed by a voivode." The word subsequently also appeared in 1886 in the sense of "the office or dignity of a voivode."[6]

An official Polish body, the Commission on Standardization of Geographic Names outside the Republic of Poland, recommends the spelling "voivodship", without the *e*.[7] This is consistently reflected in publications [8] and in the international arena, e.g., at the United Nations.[9]

See also

- List of Polish voivodeships by GDP per capita
- Voivodeships of Poland (1975–1998)
- Administrative division of the Polish–Lithuanian Commonwealth
- Coats of arms of Polish voivodeships
- Flags of Polish voivodeships
- ISO 3166-2:PL
- *Prowincja*
- Regions of Poland
- Voivodeship

Notes

[1] Also spelled "voivodship," "voi**e**vodship," "voi**e**vod**e**ship".

[2] The word "voivodeship" appears in some larger English dictionaries, such as the OED and Webster's Third New International Dictionary, but it is not in common usage. Thus, to facilitate understanding outside Poland, the word "province" is a recommended translation: "*Jednostki podziału administracyjnego Polski tłumaczymy tak: województwo*—province..." ("Polish administrative units are translated as follows: *województwo*—province..."). Arkadiusz Belczyk, *"Tłumaczenie polskich nazw geograficznych na język angielski"* (http://serwistlumacza.com/content/view/27/32/) ("Translation of Polish Geographical Names into English"), 2002-2006. Examples: New Provinces of Poland (1998) (http://www.staff.amu.edu.pl/~zbzw/ph/pro/plpro.html), Map of Poland (http://www.map-of-Poland.co.uk/), English names of Polish provinces (http://serwistlumacza.com/content/view/78/). More examples:

- "Following the reform of the administrative structure in 1973-1975, the number of **provinces** (*województwa*) was increased from 22 to 49... [I]ncreasing the number of **provinces** meant the reduction of each in size. In this way Warsaw was able to dilute the political importance of the **provincial** party chiefs." "Poland", *The Encyclopedia Americana*, 1986, volume 22, p. 312.
- "Poland is divided into 49 **provinces**." "Poland", *The Columbia Encyclopedia*, sixth edition, edited by Paul Lagassé, Columbia University Press, 2000, p. 2256.
- "Local government in Poland is organized on three levels. The largest units, at the regional level, are the *województwa* (**provinces**)..." "Poland", *Encyclopaedia Britannica*, 15th edition, 2010, *Macropaedia*, volume 25, p. 937.
- "**GOVERNMENT**... **Administrative divisions**: 16 **provinces** (wojewodztwa, singular–wojewodztwo)..." "Poland," in Central Intelligence Agency, *The CIA World Factbook 2010*, New York, Skyhorse Publishing, Inc., 2009, ISBN 978602397279 , p. 546. The same information appears in the current online *CIA World Factbook* --> "Poland --> Administrative divisions". (https://www.cia.gov/library/publications/the-world-factbook/geos/pl.html) Note that in this source, where "English translations" of province names are given, they are in the noun ("Silesia"), not the adjective ("Silesia**n**"), form.

[3] http://www.rzeczpospolita.pl/teksty/wydanie_040302/kraj_a_7-1.F.jpg

[4] http://www.gugik.gov.pl/komisja/english/pliki/podzial_administracyjny_polski_2006-eng.pdf

[5] http://www.gugik.gov.pl/komisja/english/pliki/polska_2006-podzial_administracyjny.jpg

[6] "Voivodeship," *The Oxford English Dictionary*, second edition, volume XIX, Oxford, Clarendon Press, 1989, p. 739.

[7] http://ksng.gugik.gov.pl/pliki/topon_inte.doc.pdf

[8] http://ksng.gugik.gov.pl/english/sopgn.php

[9] http://ksng.gugik.gov.pl/pliki/gegn22wp18.pdf

References

- "Poland", *Encyclopaedia Britannica*, 15th edition, 2010, *Macropaedia*, volume 25, p. 937.
- "Poland", *The Columbia Encyclopedia*, sixth edition, edited by Paul Lagassé, Columbia University Press, 2000, p. 2256.
- "Poland", *The Encyclopedia Americana*, 1986, volume 22, p. 312.
- "Poland," in Central Intelligence Agency, *The CIA World Factbook 2010*, New York, Skyhorse Publishing, Inc., 2009, ISBN 978602397279 , p. 546.
- "Voivodeship," *The Oxford English Dictionary*, second edition, volume XIX, Oxford, Clarendon Press, 1989, p. 739.

External links

- Regions of Poland (http://regiony.poland.gov.pl/start_en.html)
- Toponymic Guidelines Of Poland for Map Editors and Other Users (http://ksng.gugik.gov.pl/pliki/topon_inte.doc.pdf) Head Office Of Geodesy And Cartography, 2002
- *CIA World Factbook* --> "Poland --> Administrative divisions" (https://www.cia.gov/library/publications/the-world-factbook/geos/pl.html)

Poland

<table>
<tr><td colspan="3">Republic of Poland
Rzeczpospolita Polska</td></tr>
<tr><td colspan="3"> </td></tr>
<tr><td colspan="3">Anthem:
Mazurek Dąbrowskiego
(Poland Is Not Yet Lost)</td></tr>
<tr><td colspan="3">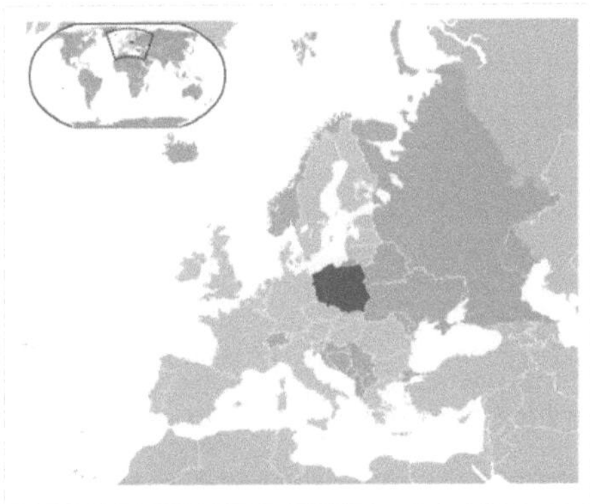
Location of Poland(dark green)
– in Europe(green & dark grey)
– in the European Union(green) — [Legend]</td></tr>
<tr><td colspan="2">Capital
and largest city</td><td>Warsaw
52°13′N 21°02′E</td></tr>
<tr><td colspan="2">Official language(s)</td><td>Polish</td></tr>
<tr><td colspan="2">Regional language(s)</td><td>Kashubian</td></tr>
<tr><td colspan="2">Ethnic groups (2011)</td><td>91.56% Polish
0.94% Silesian
0.04% Kashubian
0.07% German
0.07% Ukrainian
0.08% Belarusian
7.24% Other[1]</td></tr>
<tr><td colspan="2">Demonym</td><td>Pole/Polish</td></tr>
<tr><td colspan="2">Government</td><td>Parliamentary republic</td></tr>
<tr><td>-</td><td>President</td><td>Bronisław Komorowski</td></tr>
<tr><td>-</td><td>Prime Minister</td><td>Donald Tusk</td></tr>
<tr><td colspan="2">Legislature</td><td>National Assembly</td></tr>
</table>

-	Upper house	Senate
-	Lower house	Sejm
Formation		
-	Christianisation[c]	April 14, 966
-	Kingdom of Poland	April 18, 1025
-	Polish-Lithuanian commonwealth	July 1, 1569
-	Partition of Poland	October 24, 1795
-	Duchy of Warsaw	July 22, 1807
-	Congress Poland	June 9, 1815
-	Reconstitution of Poland	November 11, 1918
-	Invasion of Poland, World War II	September 1, 1939
-	Communist Poland	April 8, 1945
-	Republic of Poland	September 13, 1989 –
Area		
-	Total	312685 km^2 [d](69th) 120,696.41 sq mi
-	Water (%)	3.07
Population		
-	2010 estimate	38,186,860[2] (34th)
-	2011 census	38,501,000
-	Density	120/km^2 (83rd) 319.9/sq mi
GDP (PPP)		2011 estimate
-	Total	$771.658 billion[3]
-	Per capita	$20,334[3]
GDP (nominal)		2011 estimate
-	Total	$513.821 billion[3]
-	Per capita	$13,540[3]
Gini (2002)		34.5 (low
HDI (2011)		▲0.813[4] (very high / 39th)
Currency		*Złoty* (PLN)
Time zone		CET (UTC+1)
-	Summer (DST)	CEST (UTC+2)
Drives on the		right
Calling code		48
ISO 3166 code		PL
Internet TLD		.pl
1.	**a** See, however, Unofficial mottos of Poland.	

2.	**b** Although not official languages, Belarusian, Kashubian, Lithuanian and German are used in 20 communal offices.
3.	**c** The adoption of Christianity in Poland is seen by many Poles, regardless of their religious affiliation or lack thereof, as one of the most significant national historical events; the new religion was used to unify the tribes in the region.
4.	**d** The area of Poland according to the administrative division, as given by the Central Statistical Office, is 312679 km^2 (**unknown operator: u'strong'** sq mi) of which 311888 km^2 (**unknown operator: u'strong'** sq mi) is land area and 791 km^2 (**unknown operator: u'strong'** sq mi) is internal water surface area.[5]

Poland [i]/ˈpoʊlənd/ (Polish: *Polska*), officially the **Republic of Poland** (Polish: *Rzeczpospolita Polska*; Kashubian: *Pòlskô Repùblika*), is a country in Central Europe, bordered by Germany to the west; the Czech Republic and Slovakia to the south; Ukraine, Belarus and Lithuania to the east; and the Baltic Sea and Kaliningrad Oblast, a Russian exclave, to the north. The total area of Poland is 312679 square kilometres (**unknown operator: u'strong'** sq mi),[5] making it the 69th largest country in the world and the 9th largest in Europe. Poland has a population of over 38.5 million people,[5] which makes it the 34th most populous country in the world[6] and the sixth most populous member of the European Union, being its most populous post-communist member. Poland is a unitary state made up of 16 voivodeships. Poland is a member of the European Union, NATO, the United Nations, the World Trade Organization, the Organisation for Economic Co-operation and Development (OECD), European Economic Area, International Energy Agency, Council of Europe, Organization for Security and Co-operation in Europe, International Atomic Energy Agency, European Space Agency, G6, Council of the Baltic Sea States, Visegrád Group, Weimar Triangle and Schengen Agreement.

The establishment of a Polish state is often identified with the adoption of Christianity by its ruler Mieszko I in 966,[7] over the territory similar to that of present-day Poland. The Kingdom of Poland was formed in 1025, and in 1569 it cemented a long association with the Grand Duchy of Lithuania by signing the Union of Lublin, forming the Polish–Lithuanian Commonwealth. The Commonwealth ceased to exist in 1795 as the Polish lands were partitioned among the Kingdom of Prussia, the Russian Empire, and Old Austria. Poland regained its independence as the Second Polish Republic in 1918. Two decades later, in September 1939, World War II started with the Nazi Germany and Soviet Union invasion of Poland (Molotov–Ribbentrop Pact). Over six million Polish citizens died in the war. The People's Republic was declared in 1952 although Poland was a client state of the Soviet Union from 1944. During the Revolutions of 1989, the communist state was overthrown and democratic rule was re-established in the form of the current Poland, constitutionally known as the "Third Polish Republic".

Despite the vast destruction the country experienced in World War II, Poland managed to preserve much of its cultural wealth. There are currently 14 heritage sites inscribed on the UNESCO World Heritage list in Poland.[8] Since the end of the communist period, Poland has achieved a "very high" ranking in terms of human development.[9]

Etymology

The source of the name Poland[10] and the ethnonyms for the Poles[11] include endonyms (the way Polish people refer to themselves and their country) and exonyms (the way other peoples refer to the Poles and their country). Endonyms and most exonyms for Poles and Poland derive from the name of the West Slavic tribe of the Polans (*Polanie*).

The origin of the name *Polanie* itself is uncertain. It may derive from such Polish words as *pole* (field).[12] The early tribal inhabitants denominated it from the nature of the country. Lowlands and low hills predominate throughout the vast region from the Baltic shores to the foothills of the Carpathian Mountains. *Inter Alpes Huniae et Oceanum est Polonia, sic dicta in eorum idiomate quasi Campania* is the description by Gervase of Tilbury in his *Otia imperialia* (Recreation for the emperor, 1211). In some languages the exonyms for Poland derive from another tribal name, Lechites (*Lechici*).

History

Prehistory

Historians have postulated that throughout Late Antiquity, many distinct ethnic groups populated the regions of what is now known as Poland. The ethnicity and linguistic affiliation of these groups have been hotly debated; the time and route of the original settlement of Slavic peoples in these regions have been the particular subjects of much controversy.[13]

Map of Poland (Polish: Polska) in 960-992 under Mieszko, her first crowned sovereign

The most famous archeological find from the prehistory and protohistory of Poland is the Biskupin fortified settlement (now reconstructed as a museum), dating from the Lusatian culture of the early Iron Age, around 700 BC. Before adopting Christianity in 960 AD, the people of Poland believed in Svetovid, the Slavic god of war, fertility, and abundance. Many other Slavic nations had the same belief.

Piast dynasty

Baptism of Poland in 966 AD.

Poland began to form into a recognizable unitary and territorial entity around the middle of the 10th century under the Piast dynasty. Poland's first historically documented ruler, Mieszko I, was baptized in 966, adopting Catholicism as the nation's new official religion, to which the bulk of the population converted in the course of the next few centuries. In 1000 Boleslaw the Brave, continuing the policy of his father held a Congress of Gniezno and created a new dioceses (Gniezno, Kraków, Kołobrzeg, Wrocław).

In 1109 Bolesław III Wrymouth defeated in Battle of Hundsfeld King of Germany Henry V. In 1138, Poland fragmented into several smaller duchies when Bolesław III Wrymouth divided the nation amongst his sons. In 1226 Konrad I of Masovia, one of the regional Piast dukes, invited the Teutonic Knights to help him fight the Baltic Prussian pagans; a decision which would ultimately lead to centuries of warfare with the Knights. In the middle of 13th-century Poland was almost united by Silesian branch of Piast dynasty (Henry I the Bearded and Henry II the Pious), when the country was devastated by the Mongols and the Battle of Legnica where Duke Henry II the Pious died. In 1320, after a number of earlier unsuccessful attempts by regional rulers at uniting the Polish dukedoms, Władysław I consolidated his power, took the throne and became the first King of a reunified Poland. His son, Casimir III, is remembered as one of the greatest Polish kings, who was widely recognized as a protector of trade. He extended his kingdom to 250% of its initial size. Casimir is also known for extending royal protection to Jews and providing the original impetus for the establishment of Poland's first university. The Golden Liberty of the nobles began to develop under Casimir's rule, when in return for their military support, the king made serious concessions to the aristocrats, finally establishing their status as superior to that of the townsmen, and aiding their rise to power. When Casimir died in 1370 he left no legitimate male heir and, considering his other male descendants either too young or unsuitable, was laid to rest as the last of the nation's Piast rulers.

Poland was also a centre of migration of peoples. The Germans settled in the towns, the Jewish community began to settle and flourish in Poland during this era (see History of the Jews in Poland); the same applies in smaller number to Armenians. The Black Death which affected most parts of Europe from 1347 to 1351 affected Poland to lesser extent.[14]

Jagiellon dynasty

Nicolaus Copernicus 'talking with God', who reveals the truth about the solar system to him. Painting by Jan Matejko.

The rule of the Jagiellon dynasty spanned the late Middle Ages and early Modern Era of Polish history. Beginning with the Lithuanian Grand Duke Jogaila (Władysław II Jagiełło), the Jagiellon dynasty (1386–1572) formed the Polish–Lithuanian union. The partnership brought vast Lithuania-controlled Rus' areas into Poland's sphere of influence and proved beneficial for the Poles and Lithuanians, who coexisted and cooperated in one of the largest political entities in Europe for the next four centuries. In the Baltic Sea region Poland's struggle with the Teutonic Knights continued and included the Battle of Grunwald (1410), where a Polish-Lithuanian army inflicted a decisive defeat on the Teutonic Knights, both countries' main adversary, allowing Poland's and Lithuania's territorial expansion into the far north region of Livonia.[15] In 1466, after the Thirteen Years' War, King Casimir IV Jagiellon gave royal consent to the milestone Peace of Thorn, which created the future Duchy of Prussia, a Polish vassal. The Jagiellons at one point also established dynastic control over the kingdoms of Bohemia (1471 onwards) and Hungary.[16] [17] In the south Poland confronted the Ottoman Empire and the Crimean Tatars (by whom they were attacked on 75 separate occasions between 1474 and 1569),[18] and in the east helped Lithuania fight the Grand Duchy of Moscow. Some historians estimate that Crimean Tatar slave-raiding cost Poland one million of its population from 1494 to 1694.[19]

Wawel, the seat of Polish kings. Kraków was the nation's capital from 1038 until the move to Warsaw in 1596

Poland was developing as a feudal state, with a predominantly agricultural economy and an increasingly powerful landed nobility. The *Nihil novi* act adopted by the Polish Sejm (parliament) in 1505, transferred most of the legislative power from the monarch to the Sejm, an event which marked the beginning of the period known as "Golden Liberty", when the state was ruled by the "free and equal" Polish nobility. Protestant Reformation movements made deep inroads into Polish Christianity, which resulted in the establishment of policies promoting religious tolerance, unique in Europe at that time. It is believed that this tolerance allowed the country to avoid the religious turmoil that spread over Europe during the late Middle Ages. The European Renaissance evoked in late Jagiellon Poland (kings Sigismund I the Old and Sigismund II Augustus) a sense of urgency in the need to promote a cultural awakening, and resultantly during this period Polish culture and the nation's economy flourished. In 1543 the Pole, Nicolaus Copernicus, an astronomer from Toruń, published his epochal works, *De revolutionibus orbium coelestium* (*On the Revolutions of the Celestial Spheres*), and thus became the first proponent of a predictive mathematical model confirming heliocentric theory which ultimately became the accepted basic model for the practice of modern astronomy. Another major figure associated with the era is classicist poet Jan Kochanowski.[20]

Polish–Lithuanian Commonwealth

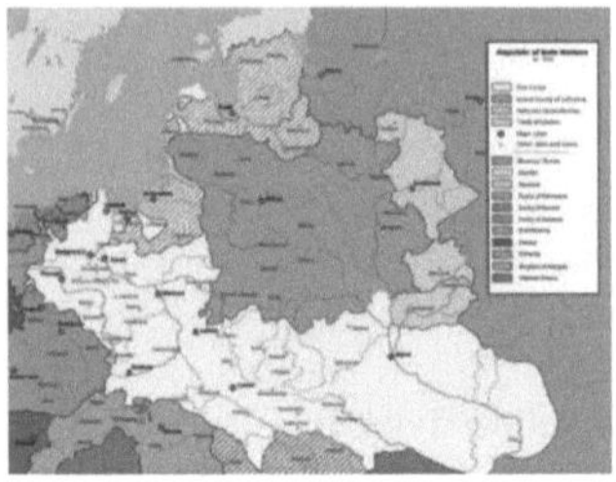

The Polish–Lithuanian Commonwealth at its greatest extent, after the Truce of Deulino (Dywilino) of 1619

The 1569 Union of Lublin established the Polish–Lithuanian Commonwealth, a more closely unified federal state with an elective monarchy, but which was governed largely by the nobility, through a system of local assemblies with a central parliament. The establishment of the Commonwealth coincided with a period of great stability and prosperity in Poland, with the union soon thereafter becoming a great European power and a major cultural entity, occupying approximately one million square kilometres of central Europe, as well as an agent for the of the 'Western culture' through Polonization in modern-day Ukraine, Belarus and Western Russia. Poland-Lithuania suffered from a number of dynastic crises during the reigns of the Vasa kings Sigismund III and Władysław IV and found itself engaged in a major conflicts with Russia, Sweden and the Ottoman Empire, as well as a series of minor Cossack uprisings.[21]

A stylised depiction of King John III Sobieski, allied commander-in-chief and victor at the Battle of Vienna

From the middle of the 17th century, the nobles' democracy, suffering from internal disorder, gradually declined, thus leaving the once powerful Commonwealth extremely vulnerable to foreign intervention. From 1648, the Cossack Khmelnytsky Uprising engulfed the south and east eventually leaving Ukraine divided, with the eastern part, lost by the Commonwealth, becoming a dependency of the Tsardom of Russia. This was soon followed by the 'Deluge', a Swedish invasion, which raged through the Polish heartlands and caused unprecedented damage to Poland's population, culture and infrastructure. Famines and epidemics followed hostilities, and the population dropped from roughly 11 to 7 million.[22] However, under John III Sobieski the Commonwealth's military prowess was re-established, and in 1683 Polish forces played a major part in relieving Vienna of a major Turkish siege which was being conducted by Kara Mustafa in hope of eventually marching his troops further into Europe to spread Islam. Unfortunately, Sobieski's reign was to mark the end of the nation's golden-era, and soon, finding itself subjected to almost constant warfare and suffering enormous population losses as well as massive damage to its economy, the Commonwealth fell into decline. The government became ineffective as a result of large scale internal conflicts (e.g. Lubomirski's Rokosz against John II Casimir and rebellious confederations) and corrupted legislative processes. The nobility fell under the control of a handful of *magnats*, and this, compounded with two relatively weak kings of the Saxon Wettin dynasty, Augustus II and Augustus III, as well as the rise of Russia and Prussia after the Great Northern War only served to worsen the Commonwealth's plight. Despite this The Commonwealth-Saxony personal union gave rise to the emergence of the Commonwealth's first reform movement, and laid the foundations for the Polish Enlightenment.[23]

During the later part of the 18th century, the Commonwealth made attempts to implement fundamental internal reforms; with the second half of the century bringing a much improved economy, significant population growth and far-reaching progress in the areas of education, intellectual life, art, and especially toward the end of the period, evolution of the social and political system. The most populous capital city of Warsaw replaced Gdańsk (Danzig) as the leading centre of commerce, and the role of the more prosperous townsfolk soon increased. The royal election of 1764 resulted in the elevation of Stanisław August Poniatowski, a refined and worldly aristocrat connected to a major magnate faction, to the monarchy. However, a one-time lover of Empress Catherine II of Russia, the new king spent much of his reign torn between his desire to implement reforms necessary to save his nation, and his perceived necessity to remain in a relationship with his Russian sponsor. This ultimately led to the formation of the 1768 Bar Confederation; a *szlachta* rebellion directed against Russia and the Polish king which fought to preserve Poland's independence and the *szlachta*'s traditional privileges. Attempts at reform provoked the union's neighbours, and in 1772 the First Partition of the Commonwealth by Russia, Austria and Prussia took place; an act which the "Partition Sejm", under considerable duress, eventually "ratified" *fait accompli*.[24] Disregarding this loss, in 1773 the king established the Commission of National Education, the first government education authority in Europe.

Stanisław August Poniatowski, the last King of Poland and Grand Duke of Lithuania acceded to the throne in 1764, reigning until his abdication on 25 November 1795

The long-lasting Great Sejm convened by Stanisław August in 1788 successfully adopted the May 3 Constitution, the first set of modern supreme national laws in Europe. However, this document, accused by detractors of harbouring revolutionary sympathies, soon generated strong opposition from the Commonwealth's nobles and conservatives as well as from Catherine II, who, determined to prevent the rebirth of a strong Commonwealth set about planning the final dismemberment of the Polish-Lithuanian state. Russia was greatly aided in achieving its goal when the Targowica Confederation, an organisation of Polish nobles, appealed to the Empress for help, and in May 1792 Russian forces crossed the Commonwealth's frontier, thus beginning the Polish-Russian War. The defensive war fought by the Poles and Lithuanians ended prematurely when the King, convinced of the futility of resistance, capitulated and joined the Targowica Confederation. The Confederation then took over the government; Russia and Prussia, fearing the mere existence of a Polish state, arranged for, and subsequently in 1793, executed the Second Partition of the Commonwealth, which left the country deprived of so much territory that it was practically incapable of independent existence. Eventually, in 1795, following the failed Kościuszko Uprising, the Commonwealth was partitioned one last time by all three of its more powerful neighbours, and with this, effectively ceased to exist.[25]

The Age of Partitions

Tadeusz Kościuszko takes the oath to the King on the Rynek in Kraków, 1794

Poles rebelled several times against the partitioners, particularly near the end of the 18th century and the beginning of the 19th century. One of the most famous and successful attempts at securing renewed Polish independence took place in 1794, during the Kościuszko Uprising, at the Racławice where Tadeusz Kosciuszko, a popular and distinguished general who had served under Washington in America, led peasants and some Polish regulars into battle against numerically superior Russian forces. In 1807, Napoleon I of France recreated a Polish state, the Duchy of Warsaw, but after the Napoleonic Wars, Poland was again divided by the victorious Allies at the Congress of Vienna of 1815. The eastern part was ruled by the Russian tsar as a Congress Kingdom which possessed a very liberal constitution. However, the tsars soon reduced Polish freedoms, and Russia annexed the country in virtually all but name. Thus in the latter half of the 19th century, only Austrian-ruled Galicia, and particularly the Free City of Kraków, created good environment for free Polish cultural life to flourish.

Throughout the period of the partitions, political and cultural repression of the Polish nation led to the organisation of a number of uprisings against the authorities of the occupying Russian, Prussian and Austrian governments. Notable amongst these are the November Uprising of 1830 and January Uprising of 1863, both of which were attempts to free Poland from the rule of tsarist Russia. The November uprising began on 29 November 1830 in Warsaw when, led by Lieutenant Piotr Wysocki, young non-commissioned officers at the Imperial Russian Army's military academy in that city revolted. They were soon joined by large segments of Polish society, and together forced Warsaw's Russian garrison to withdraw north of the city.

Polish insurgents and Russian cuirassiers clash on a bridge in Warsaw's Łazienki Park during the November Uprising, Painting by Wojciech Kossak, 1898

Over the course of the next seven months, Polish forces successfully defeated the Russian armies of Field Marshal Hans Karl von Diebitsch and a number of other Russian commanders; however, finding themselves in a position unsupported by any other foreign powers, save distant France and the newborn United States, and with Prussia and Austria refusing to allow the import of military supplies through their territories, the Poles accepted that the uprising was doomed to failure. Upon the surrender of Warsaw to General Ivan Paskievich, many Polish troops, feeling they could not go on, withdrew into Germany and there laid down their arms. Poles would have to wait another 32 years for another opportunity to free their homeland.

When in January 1863 a new Polish uprising against Russian rule began, it did so as a spontaneous protest by young Poles against conscription into the Imperial Russian Army. However, the insurrectionists, despite being joined by high-ranking Polish-Lithuanian officers and numerous politicians were still severely outnumbered and lacking in foreign support. They were forced to resort to guerrilla warfare tactics and ultimately failed to win any major military victories. Afterwards no major uprising was witnessed in the Russian controlled Congress Poland and Poles resorted instead to fostering economic and cultural self-improvement.

Despite the political unrest experienced during the partitions, Poland did benefit from large scale industrialisation and modernisation programs, instituted by the occupying powers, which helped it develop into a more economically coherent and viable entity. This was particularly true in the Greater Poland, Pomerania and Warmia annexed by Prussia (later becoming a part of the German Empire); an area which eventually, thanks largely to the Greater Poland

Uprising, was reconstituted as a part of the Second Polish Republic and became one of its most productive regions.

Reconstitution of Poland

Chief of State Marshal Józef Piłsudski

During World War I, all the Allies agreed on the reconstitution of Poland that United States President Woodrow Wilson proclaimed in Point 13 of his Fourteen Points. A total of 2 million Polish troops fought with the armies of the three occupying powers, and 450,000 died.[26] Shortly after the armistice with Germany in November 1918, Poland regained its independence as the Second Polish Republic (*II Rzeczpospolita Polska*). It reaffirmed its independence after a series of military conflicts, the most notable being the Polish–Soviet War (1919–1921) when Poland inflicted a crushing defeat on the Red Army at the Battle of Warsaw, an event which is considered to have ultimately halted the advance of Communism into Europe and forced Vladimir Lenin to rethink his objective of achieving global socialism. Nowadays the event is often referred to as the 'Miracle at the Vistula'.[27]

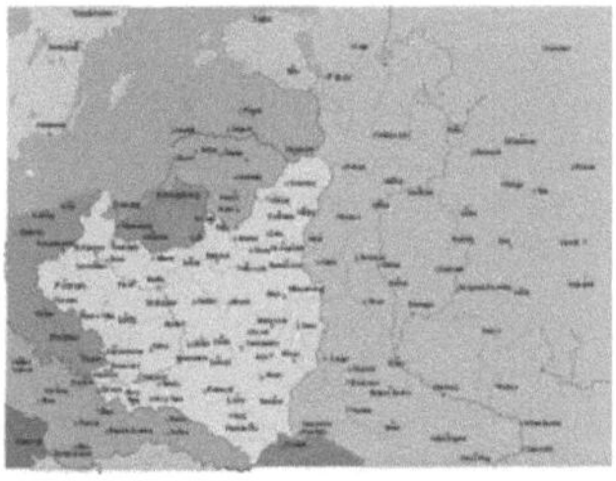

Poland between 1922 and 1938

During this period, Poland successfully managed to fuse the territories of the three former partitioning powers into a cohesive nation state. Railways were restructured to direct traffic towards Warsaw instead of the former imperial capitals, a new network of national roads was gradually built up and a major seaport was opened on the Baltic Coast, so as to allow Polish exports and imports to bypass the politically charged Free City of Danzig.

The inter-war period heralded in a new era of Polish politics. Whilst Polish political activists had faced heavy censorship in the decades up until the First World War, the country now found itself trying to establish a new political tradition. For this reason, many exiled Polish activists, such as Jan Paderewski (who would later become Prime Minister) returned home to help; a great number of them then went on to take key positions in the newly formed political and governmental structures. Tragedy struck in 1922 when Gabriel Narutowicz, inaugural holder of the Presidency, was assassinated at the Zachęta Gallery in Warsaw by painter and right-wing nationalist Eligiusz Niewiadomski.[28]

The 1926 May Coup of Józef Piłsudski turned rule of the Second Polish Republic over to the Sanacja movement. By the 1930s Poland had become increasingly authoritarian; a number of 'undesirable' political parties, such as the Polish Communists, had been banned and following Piłsudski's death, the regime, unable to appoint a new leader, began to show its inherent internal weaknesses and unwillingness to cooperate in any way with other political parties.

World War II

The Sanacja movement controlled Poland until the start of World War II in 1939, when Nazi Germany's and Slovakia Invasion of Poland (1939) on 1 September and the Soviet invasion of Poland on 17 September, which followed the breaking of the Soviet–Polish Non-Aggression Pact, occurred. Warsaw capitulated on 28 September 1939. As agreed in the Molotov–Ribbentrop Pact, Poland was split into two zones, one occupied by Nazi Germany while the Kresy, or Borderlands, fell under the control of the Soviet Union. In 1939-1941, the Soviets had moved hundreds of thousands of Poles across the Soviet Union, and the Soviet secret police, NKVD, had executed thousands of Polish prisoners of war (inter alia Katyn massacre).[29]

Poland made the fourth-largest troop contribution to the Allied war effort, after the Soviets, the British and the Americans.[a] Polish troops fought under the command of both the Polish Government in Exile in the theatre of war west of Germany and under Soviet leadership in the theatre of war east of Germany. The Polish expeditionary corps, which was controlled by the exiled pre-war government based in London, played an important role in the Italian and North African Campaigns.[30] [31] They are particularly well remembered for their conduct at the Battle of Monte Cassino, a conflict which culminated in the raising of a Polish flag over the ruins of the mountain-top abbey by the 12th Podolian Uhlans. The Polish forces in the theatre of war east of Germany were commanded by Lieutenant General Władysław Anders who had received his command from Prime Minister of the exiled government Władysław Sikorski. On the east of Germany, the Soviet-backed Polish 1st Army distinguished itself in the battles for Berlin and Warsaw, although its actions in support of the latter have often been criticised.

Polish forces stationed abroad constituted the fourth largest allied force of the war

At the end of World War II, the gray territories were transferred from Poland to the Soviet Union, and the pink territories from Germany to Poland

Polish servicemen were also active in the theatres of naval and air warfare; during the Battle of Britain Polish squadrons such as the No. 303 "Kościuszko" fighter squadron[32] achieved great success, and by the end of the war the exiled Polish Air Forces could claim 769 confirmed kills. Meanwhile, the Polish Navy was active in the protection of convoys in the North Sea and Atlantic Ocean.[33]

In addition to the organised units of the 1st Army and the Forces in the Nazi-occupied Europe, the domestic underground resistance movement, the Armia Krajowa, or *Home Army*, fought to free Poland from German occupation and establish an independent Polish state. The wartime resistance movement in Poland was one of the three largest resistance movements of the entire war[b] and encompassed an unusually broad range of clandestine activities, which essentially functioned as an underground state complete with degree-awarding universities and a court system.[34] The resistance was, however, largely loyal to the exiled government and generally resented the idea of a communist Poland; for this reason, on 1 August 1944 they initiated Operation Tempest and thus began the Warsaw Uprising.[35] [36] The objective of the uprising was to drive the German occupiers from the city and help with the larger fight against Germany and the Axis powers, however secondary motives for the uprising sought to see Warsaw liberated before the Soviets could reach the capital, so as to underscore Polish sovereignty by empowering the Polish Underground State before the Soviet-backed Polish Committee of National Liberation could assume control. However, a lack of available allied military aid and Stalin's reluctance to allow the 1st Army to help their fellow countrymen take the city, ultimately led to the uprising's failure and subsequent planned destruction of the city.

Grave of Polish fighter killed during the Warsaw Uprising. The battle lasted 63 days, and resulted in the deaths of 200,000 people.

During the war, German forces, under direct order from Adolf Hitler, set up six major extermination camps, all of which were established on Polish territory; these included both the notorious Treblinka and Auschwitz camps. This allowed the Germans to transport the Jews living in the Third Reich outside of "German" territory, as well as to import Jews and other targeted groups from across occupied Europe to be "liquidated," or killed, in the concentration camps set up in the General Government. Among such groups were Polish intelligentsia, communists, Roma peoples and Soviet Prisoners of War. However, since millions of Jews lived in pre-war Poland, Jewish victims make up the largest percentage of all victims of the Nazis' extermination program. It is estimated that, of pre-war Poland's Jewry, approximately 90% (or about 3 million members) were killed. Throughout the occupation, many members of the Armia Krajowa, supported by the Polish government in exile, and millions of ordinary Poles — at great risk to themselves and their families — engaged in rescuing Jews from the Nazi Germans. Grouped by nationality, Poles represent the largest number of people who rescued Jews during the Holocaust.[37] [38] To date, 6,135 Poles have been awarded the title of *Righteous among the Nations* by the State of Israel–more than any other nation.[37] Some estimates put the number of Poles involved in rescue efforts at up to 3 million, and credit Poles with saving up to approximately 450,000 Jews from certain death.[38]

At the war's conclusion, Poland's territory was shifted westwards, pushing the Kresy in accordance with the Curzon Line. Meanwhile, the western border moved to the Oder-Neisse line. As a result, Poland's territory was reduced by 20%, or 77500 square kilometres (**unknown operator: u'strong'** sq mi). The shift forced the migration of millions of people, most of whom were Poles, Germans, Ukrainians, and Jews.[39] Of all the countries involved in the war, Poland lost the highest percentage of its citizens: over 6 million perished — nearly one-fifth of Poland's population — half of them Polish Jews. Over 90% of deaths were non-military in nature. Only in the 1970s did Poland again approach its prewar population levels.

Postwar communist Poland

At the insistence of Joseph Stalin, the Yalta Conference sanctioned the formation of a new Polish provisional and pro-Communist coalition government in Moscow, which ignored the Polish government-in-exile based in London; a move which angered many Poles who considered it a betrayal by the Allies. In 1944, Stalin had made guarantees to Churchill and Roosevelt that he would maintain Poland's sovereignty and allow democratic elections to take place; however, upon achieving victory in 1945, the occupying Soviet authorities organised an election which constituted nothing more than a sham and was ultimately used to claim the 'legitimacy' of Soviet hegemony over Polish affairs. The Soviet Union instituted a new communist government in Poland, analogous to much of the rest of the Eastern Bloc. As elsewhere in Communist Europe the Soviet occupation of Poland met with armed resistance from the outset which continued into the fifties.

At High Noon, June 4, 1989 – political poster featuring Gary Cooper to encourage votes for the Solidarity (Polish trade union)Solidarity party in the Contract Sejm1989 elections. "The 4th of June, 1989 marked a decisive victory for democracy in Poland and, ultimately, across (Central and) Eastern Europe.— Angela Merkel (English) "Merkel honours Polish freedom struggle and Tiananmen victims". www.thelocal.de. AFP. . Retrieved 2009-09-28.

Despite widespread objections, the new Polish government accepted the Soviet annexation of the pre-war eastern regions of Poland[41] (in particular the cities of Wilno and Lwów) and agreed to the permanent garrisoning of Red Army units on Poland's territory. Military alignment within the Warsaw Pact throughout the Cold War came about as a direct result of this change in Poland's political culture and in the European scene came to characterise the fully-fledged integration of Poland into the brotherhood of communist nations.

The People's Republic of Poland (*Polska Rzeczpospolita Ludowa*) was officially proclaimed in 1952. In 1956 after the death of Bolesław Bierut, the régime of Władysław Gomułka became temporarily more liberal, freeing many people from prison and expanding some personal freedoms. A similar situation repeated itself in the 1970s under Edward Gierek, but most of the time persecution of anti-communist opposition groups persisted. Despite this, Poland was at the time considered to be one of the least oppressive states of the Soviet Bloc.[42]

Gomułka with Leonid Brezhnev in East Germany

Labour turmoil in 1980 led to the formation of the independent trade union "Solidarity" ("*Solidarność*"), which over time became a political force. Despite persecution and imposition of martial law in 1981, it eroded the dominance of the Communist Party and by 1989 had triumphed in Poland's first partially free and democratic parliamentary elections since the end of the Second World War. Lech Wałęsa, a Solidarity candidate, eventually won the presidency in 1990. The Solidarity movement heralded the collapse of communist regimes and parties across Europe.

Present-day Poland

Poland joined NATO in 1999 and since 2004 has been a member of the European Union.

A shock therapy programme, initiated by Leszek Balcerowicz in the early 1990s enabled the country to transform its socialist-style planned economy into a market economy. As with all other post-communist countries, Poland suffered temporary slumps in social and economic standards, but it became the first post-communist country to reach its pre-1989 GDP levels, which it achieved by 1995 largely thanks to its booming economy.[43] [44]

Most visibly, there were numerous improvements in human rights, such as the freedom of speech, civil liberties (1st class) and political rights (1st class), according to Freedom House. In 1991, Poland became a member of the Visegrád Group and joined the North Atlantic Treaty Organization (NATO) alliance in 1999 along with the Czech Republic and Hungary. Poles then voted to join the European Union in a referendum in June 2003, with Poland becoming a full member on 1 May 2004. Subsequently Poland joined the Schengen Area in 2007, as a result of which, the country's borders with other member states of the European Union have been dismantled, allowing for full freedom of movement within most of the EU.[45] In contrast to this, the section of Poland's eastern border now comprising the external EU border with Belarus, Russia and Ukraine, has become increasingly well protected, and has led in part to the coining of the phrase 'Fortress Europe', in reference to the seeming 'impossibility' of gaining entry to the EU for citizens of the former Soviet Union.

On April 10, 2010, the President of the Republic of Poland, Lech Kaczyński, along with 89 other high-ranking Polish officials died in a plane crash near Smolensk, Russia. The president's party were on their way to attend an annual service of commemoration for the victims of the Katyń massacre when the tragedy took place.

Geography

Poland's topography

Poland's territory extends across several geographical regions, between latitudes 49° and 55° N, and longitudes 14° and 25° E. In the north-west is the Baltic seacoast, which extends from the Bay of Pomerania to the Gulf of Gdańsk. This coast is marked by several spits, coastal lakes (former bays that have been cut off from the sea), and dunes. The largely straight coastline is indented by the Szczecin Lagoon, the Bay of Puck, and the Vistula Lagoon. The centre and parts of the north lie within the North European Plain.

Rising gently above these lowlands is a geographical region comprising the four hilly districts of moraines and moraine-dammed lakes formed during and after the Pleistocene ice age. These lake districts are the Pomeranian Lake District, the Greater Polish Lake District, the Kashubian Lake District, and the Masurian Lake District. The Masurian Lake District is the largest of the four and covers much of north-eastern Poland. The lake districts form part of the Baltic Ridge, a series of moraine belts along the southern shore of the Baltic Sea.

South of the Northern European Lowlands lie the regions of Silesia and Masovia, which are marked by broad ice-age river valleys. Farther south lies the Polish mountain region, including the Sudetes, the Cracow-Częstochowa Upland, the Świętokrzyskie Mountains, and the Carpathian Mountains, including the Beskids. The highest part of the Carpathians is the Tatra Mountains, along Poland's southern border.

Geology

Granite outcrop Silesian Rocks at Karkonosze (Giant Mountains), south-western Poland

The geological structure of Poland has been shaped by the continental collision of Europe and Africa over the past 60 million years, on the one hand (and the other), by the Quaternary glaciations of northern Europe. Both processes shaped the Sudetes and the Carpathian Mountains. The moraine landscape of northern Poland contains soils made up mostly of sand or loam, while the ice age river valleys of the south often contain loess. The Cracow-Częstochowa Upland, the Pieniny, and the Western Tatras consist of limestone, while the High Tatras, the Beskids, and the Karkonosze are made up mainly of granite and basalts. The Polish Jura Chain is one of the oldest mountain ranges on earth.

Poland has 70 mountains over 2,000 metres (6,600 ft) in elevation, all in the Tatras. The Polish Tatras, which consist of the High Tatras and the Western Tatras, is the highest mountain group of Poland and of the entire Carpathian range. In the High Tatras lies Poland's highest point, the north-western peak of Rysy, 2499 metres (8199 ft) in elevation. At its foot lies the mountain lakes of Czarny Staw pod Rysami (Black Lake below Mount Rysy), and Morskie Oko (the Marine Eye).

Giewont in the Tatra Mountains; the mountainous south is a popular destination for hikers

The second highest mountain group in Poland is the Beskids, whose highest peak is Babia Góra, at 1725 metres (5659 ft). The next highest mountain groups is the Karkonosze in the Sudetes, whose highest point is Śnieżka, at 1602 metres (5256 ft); Śnieżnik Mountains whose highest point is Śnieżnik, at 1425 metres (4675 ft).

Tourists also frequent the Bieszczady Mountains in the far southeast of Poland, whose highest point in Poland is Tarnica, with an elevation of 1346 metres (4416 ft), Gorce Mountains in Gorce National Park, whose highest point is Turbacz, with elevations 1310 metres (4298 ft), and the Pieniny in Pieniny National Park, whose highest point is Wysokie Skałki (Wysoka), with elevations 1050 metres (3445 ft). The lowest point in Poland — at 2 metres (6.6 ft) below sea level — is at Raczki Elbląskie, near Elbląg in the Vistula Delta.

Dunes in Słowiński National Park

The only desert located in Poland stretches over the Zagłębie Dąbrowskie (the Coal Fields of Dąbrowa) region. It is called the Błędów Desert, located in the Silesian Voivodeship in southern Poland. It has a total area of 32 square kilometres (12 sq mi). It is one of only five natural deserts in Europe. But also, it is the warmest desert that appears at this latitude. Błędów Desert was created thousands of years ago by a melting glacier. The specific geological structure has been of big importance. The average thickness of the sand layer is about 40 metres (131 ft), with a maximum of 70 metres (230 ft), which made the fast and deep drainage very easy.

The Baltic Sea activity in Słowiński National Park created sand dunes which in the course of time separated the bay from the sea. As waves and wind carry sand inland the dunes slowly move, at a speed of 3 to 10 metres (9.8 to 32.8 ft) meters per year. Some dunes are quite high – up to 30 metres (98 ft). The highest peak of the park — Rowokol (115 metres / 377 feet above sea level) — is also an excellent observation point.

Waters

Vistula River in Modlin

The longest rivers are the Vistula (Polish: *Wisła*), 1047 kilometres (**unknown operator: u'strong'** mi) long; the Oder (Polish: *Odra*) which forms part of Poland's western border, 854 kilometres (**unknown operator: u'strong'** mi) long; its tributary, the Warta, 808 kilometres (**unknown operator: u'strong'** mi) long; and the Bug, a tributary of the Vistula, 772 kilometres (**unknown operator: u'strong'** mi) long. The Vistula and the Oder flow into the Baltic Sea, as do numerous smaller rivers in Pomerania.

The Łyna and the Angrapa flow by way of the Pregolya to the Baltic, and the Czarna Hańcza flows into the Baltic through the Neman. While the great majority of Poland's rivers drain into the Baltic Sea, Poland's Beskids are the source of some of the upper tributaries of the Orava, which flows via the Váh and the Danube to the Black Sea. The eastern Beskids are also the source of some streams that drain through the Dniester to the Black Sea.

Poland's rivers have been used since early times for navigation. The Vikings, for example, traveled up the Vistula and the Oder in their longships. In the Middle Ages and in early modern times, when the Polish–Lithuanian Commonwealth was the breadbasket of Europe; the shipment of grain and other agricultural products down the Vistula toward Gdańsk and onward to other parts of Europe took on great importance.

Kurtkowiec, oligotrophic lake in southeastern Poland

With almost ten thousand closed bodies of water covering more than 1 hectare (2.47 acres) each, Poland has one of the highest numbers of lakes in the world. In Europe, only Finland has a greater density of lakes. The largest lakes, covering more than 100 square kilometres (39 sq mi), are Lake Śniardwy and Lake Mamry in Masuria, and Lake Łebsko and Lake Drawsko in Pomerania.

In addition to the lake districts in the north (in Masuria, Pomerania, Kashubia, Lubuskie, and Greater Poland), there is also a large number of mountain lakes in the Tatras, of which the Morskie Oko is the largest in area. The lake with the greatest depth—of more than 100 metres (328 ft)—is Lake Hańcza in the Wigry Lake District, east of Masuria in Podlaskie Voivodeship.

Holidaymakers relax at the Lake Solina near Lesko in southeastern Poland

Among the first lakes whose shores were settled are those in the Greater Polish Lake District. The stilt house settlement of Biskupin, occupied by more than one thousand residents, was founded before the 7th century BC by people of the Lusatian culture.

Lakes have always played an important role in Polish history and continue to be of great importance to today's modern Polish society. The ancestors of today's Poles, the Polanie, built their first fortresses on islands in these lakes. The legendary Prince Popiel is supposed to have ruled from Kruszwica on Lake Gopło. The first historically documented ruler of Poland, Duke Mieszko I, had his palace on an island in the Warta River in Poznań. Nowadays the Polish lakes provide an invaluable location for the pursuit of water sports such as yachting and wind-surfing.

The Polish Baltic coast is approximately 528 kilometres (328 mi) long and extends from Świnoujście on the islands of Usedom and Wolin in the west to Krynica Morska on the Vistula Spit in the east. For the most part, Poland has a smooth coastline, which has been shaped by the continual movement of sand by currents and winds. This continual erosion and deposition has formed cliffs, dunes, and spits, many of which have migrated landwards to close off former lagoons, such as Łebsko Lake in Słowiński National Park.

Bay of Puck *(Zatoka Pucka)* in Poland

Prior to the end of the Second World War and subsequent change in national borders, Poland had only a very small coastline; this was situated at the end of the 'Polish Corridor', the only internationally recognised Polish territory which afforded the country access to the sea. However after World War II, the redrawing of Poland's borders and resulting 'shift' of the country's borders left it with a greatly expanded coastline, thus allowing for far greater access to the sea than was ever previously possible. The significance of this event, and importance of it to Poland's future as a major industrialised nation, was allured to by the 1945 Wedding to the Sea.

The largest spits are Hel Peninsula and the Vistula Spit. The largest Polish Baltic island is Wolin. The largest port cities are Gdynia, Gdańsk, Szczecin, and Świnoujście. The main coastal resorts are Sopot, Międzyzdroje, Kołobrzeg, Łeba, Władysławowo, and the Hel Peninsula.

Land use

Forests of the Ojców National Park

Poland is the fourth most forested country in Europe. Forests cover about 30,5% of Poland's land area based on international standards. Its overall percentage is still increasing. Forests of Poland is managed by the national program of reforestation (KPZL), aiming at an increase of forest-cover to 33% in 2050. The richness of Polish forest (per SoEF 2011 statistics) is more than twice as high as European average (with Germany and France at the top), containing 2.304 billion cubic metres of trees.[46] The largest forest complex in Poland is Lower Silesian Wilderness.

More than 1% of Poland's territory, 3145 square kilometres (**unknown operator: u'strong'** sq mi), is protected within 23 Polish national parks. Three more national parks are projected for Masuria, the Kraków-Częstochowa Upland, and the eastern Beskids. In addition, wetlands along lakes and rivers in central Poland are legally protected, as are coastal areas in the north. There are over 120 areas designated as landscape parks, along with numerous nature reserves and other protected areas (e.g. Natura 2000).

Present day Poland is a country with great agricultural prospects; there are over two million private farms in the country, and Poland is the leading producer in Europe of potatoes and rye and is one of the world's largest producers of sugar beets and triticale. This has led Poland to be described on occasion as the future 'bread basket of the European Union'. However, despite employing around 16% of the workforce, agricultural output in Poland remains low and the industry is characterised as largely inefficient due to the large number of small, independent farms. This situation is likely to soon change for the better with the government debating agricultural reform and currently pursuing the option of auctioning off large tracts of state-owned agricultural land.

Biodiversity

Family of White Stork. Poland hosts the largest White Stork population.[47]

Phytogeographically, Poland belongs to the Central European province of the Circumboreal Region within the Boreal Kingdom. According to the World Wide Fund for Nature, the territory of Poland belongs to three Palearctic Ecoregions of the continental forest spanning Central and Northern European temperate broadleaf and mixed forest ecoregions as well as the Carpathian montane conifer forest.

A wisent in the Białowieża Forest

Many animals that have since died out in other parts of Europe still survive in Poland, such as the wisent in the ancient woodland of the Białowieża Forest and in Podlaskie. Other such species include the brown bear in Białowieża, in the Tatras, and in the Beskids, the gray wolf and the Eurasian Lynx in various forests, the moose in northern Poland, and the beaver in Masuria, Pomerania, and Podlaskie.

In the forests, one also encounters game animals, such as red deer, roe deer and wild boars. In eastern Poland there are a number of ancient woodlands, like Białowieża forest, that have never been cleared by people. There are also large forested areas in the mountains, Masuria, Pomerania, Lubusz Land and Lower Silesia.

Poland is the most important breeding ground for European migratory birds. Out of all of the migratory birds who come to Europe for the summer, one quarter breed in Poland, particularly in the lake districts and the wetlands along the Biebrza, the Narew, and the Warta, which are part of nature reserves or national parks.

Climate

The average daytime summer temperature at sea level along the Baltic coast is 22 °C (71.6 °F)[48]

The climate is mostly temperate throughout the country. The climate is oceanic in the north and west and becomes gradually warmer and continental towards the south and east. Summers are generally warm, with average temperatures between 18 °C (**unknown operator: u'strong'** °F) and 30 °C (86.0 °F) depending on a region. Winters are rather cold, with average temperatures around 3 °C (37.4 °F) in the northwest and −6 °C (**unknown operator: u'strong'** °F) in the northeast. Precipitation falls throughout the year, although, especially in the east; winter is drier than summer.

The warmest region in Poland is Lower Silesian located in south-western Poland where temperatures in the summer average between 24 °C (75.2 °F) and 32 °C (**unknown operator: u'strong'** °F) but can go as high as 34 °C (93.2 °F) to 39 °C (102.2 °F) on some days in the warmest month of July and August. The warmest cities in Poland are Tarnów, which is situated in Lesser Poland and Wrocław, which is located in Lower Silesian. The average temperatures in Wrocław are 20 °C (**unknown operator: u'strong'** °F) in the summer and 0 °C (32.0 °F) in the winter, but Tarnów has the longest summer in all of Poland, which lasts for 115 days, from mid-May to mid-September. The coldest region of Poland is in the northeast in the Podlaskie Voivodeship near the border of Belarus. Usually the coldest city is Suwałki. The climate is affected by cold fronts which come from Scandinavia and Siberia. The average temperature in the winter in Podlaskie ranges from –6 °C (**unknown operator: u'strong'** °F) to –4 °C (**unknown operator: u'strong'** °F).

Politics

President of Poland Bronisław Komorowski

Poland is a democracy, with a president as a head of state, whose current constitution dates from 1997. The government structure centers on the Council of Ministers, led by a prime minister. The president appoints the cabinet according to the proposals of the prime minister, typically from the majority coalition in the Sejm. The president is elected by popular vote every five years. The president is Bronisław Komorowski. Komorowski replaced President Lech Kaczyński following the latter's death in an April 10, 2010 air crash. The prime minister, Donald Tusk, was appointed in 2007 after his Civic Platform party made significant gains in that year's parliamentary elections. In 2011, Tusk became the first Polish prime minister in history to be democratically re-elected for a consecutive term.

Polish voters elect a bicameral parliament consisting of a 460-member lower house (Sejm) and a 100-member Senate (Senat). The Sejm is elected under proportional representation according to the d'Hondt method, a method similar to that used in many parliamentary political systems. The Senat, on the other hand, is elected under the First-past-the-post voting method, with one senator being returned from each of the 100 constituencies.

The Sejm's session chamber in Warsaw

With the exception of ethnic minority parties, only candidates of political parties receiving at least 5% of the total national vote can enter the Sejm. When sitting in joint session, members of the Sejm and Senat form the National Assembly (the *Zgromadzenie Narodowe*). The National Assembly is formed on three occasions: when a new President takes the oath of office; when an indictment against the President of the Republic is brought to the State Tribunal (*Trybunał Stanu*); and when a president's permanent incapacity to exercise his duties because of the state of his health is declared. To date only the first instance has occurred.

The judicial branch plays an important role in decision-making. Its major institutions include the Supreme Court of the Republic of Poland (*Sąd Najwyższy*); the Supreme Administrative Court of the Republic of Poland (*Naczelny Sąd Administracyjny*); the Constitutional Tribunal of the Republic of Poland (*Trybunał Konstytucyjny*); and the State Tribunal of the Republic of Poland (*Trybunał Stanu*). On the approval of the Senat, the Sejm also appoints the ombudsman or the Commissioner for Civil Rights Protection (*Rzecznik Praw Obywatelskich*) for a five-year term. The ombudsman has the duty of guarding the observance and implementation of the rights and liberties of Polish citizens and residents, of the law and of principles of community life and social justice.

In 2011, Poles elected Anna Grodzka as the first ever transsexual MP in European history,[49] and the second transgender MP in European history, after the Italian Vladimir Luxuria.[50]

Law

The Supreme Court building in Warsaw

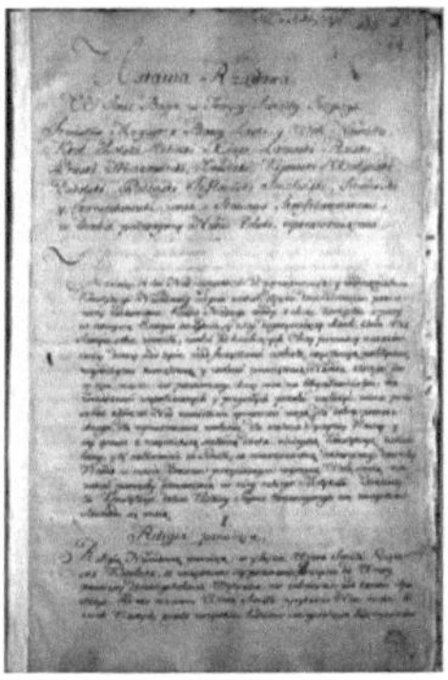

The Constitution of May 3, 1791 has been called the first of its kind in Europe.[51]

The Constitution of Poland is the supreme law in contemporary Poland, and the Polish legal system is based on the principle of civil rights, governed by the code of Civil Law. Historically, the most famous Polish legal act is the Constitution of May 3, 1791. Historian Norman Davies describes it as the first of its kind in Europe.[51] The Constitution was instituted as a Government Act (Polish: *Ustawa rządowa*) and then adopted on May 3, 1791 by the Sejm of the Polish–Lithuanian Commonwealth. Primarily, it was designed to redress long-standing political defects of the federative Polish–Lithuanian Commonwealth and its Golden Liberty. Previously only the Henrican articles signed by each of Poland's elected kings could perform the function of a set of basic laws. The new Constitution introduced political equality between townspeople and the nobility (*szlachta*), and placed the peasants under the protection of the government, thus mitigating the worst abuses of serfdom. The Constitution abolished pernicious parliamentary institutions such as the *liberum veto*, which at one time had placed the sejm at the mercy of any deputy who might choose, or be bribed by an interest or foreign power, to have rescinded all the legislation that had been passed by that sejm. The May 3rd Constitution sought to supplant the existing anarchy fostered by some of the country's reactionary magnates, with a more egalitarian and democratic constitutional monarchy.

Unfortunately, the adoption of such a liberal constitution was treated as a grave threat by Poland's more autocratic neighbours. In response Prussia, Austria and Russia formed an anti-Polish alliance and over the next decade collaborated with one another to partition their weaker neighbour and ultimately destroy the Polish state. In the words of two of its co-authors, Ignacy Potocki and Hugo Kołłątaj, the constitution represented "the last will and testament of the expiring Fatherland." Despite this, its text influenced many later democratic movements across the globe.

Poland's current constitution was adopted by the National Assembly of Poland on 2 April 1997, approved by a national referendum on 25 May 1997, and came into effect on 17 October 1997. It guarantees a multi-party state, the freedoms of religion, speech and assembly, and specifically casts off many Communist ideals to create a 'free market economic system'. It requires public officials to pursue ecologically sound public policy and acknowledges the inviolability of the home, the right to form trade unions, and to strike, whilst at the same time prohibiting the practices of forced medical experimentation, torture and corporal punishment.

Foreign relations

Foreign Secretary Sikorski meets US Secretary of State Hillary Clinton

In recent years, Poland has extended its responsibilities and position in European and international affairs, supporting and establishing friendly relations with other European nations and a large number of 'developing' countries.

In 1994, Poland became an associate member of the European Union (EU) and its defensive arm, the Western European Union (WEU), having submitted preliminary documentation for full membership in 1996, it formally joined the European Union in May 2004, along with the other members of the Visegrád group. In 1996, Poland achieved full OECD membership, and at the 1997 Madrid Summit was invited to join the North Atlantic Treaty Organisation (NATO) in the first wave of policy enlargement finally becoming a full member of NATO in March 1999.

Prime Minister Donald Tusk (right) arrives with former President Lech Wałęsa for the EPP party congress in Warsaw

As changes since the fall of Communism in 1989 have redrawn the map of central Europe, Poland has tried to forge strong and mutually beneficial relationships with its seven new neighbours, this has notably included signing 'friendship treaties' to replace links severed by the collapse of the Warsaw Pact. The Poles have forged special relationships with Lithuania and particularly Ukraine,[52] with whom they co-hosted the UEFA Euro 2012 football tournament, in an effort to firmly anchor these countries within the Western world and provide them with an alternative to aligning themselves with the Russian Federation respectively. Despite many positive developments in the region, Poland has found itself in a position where it must seek to defend the rights of ethnic Poles living in the former Soviet Union; this is particularly true of Belarus, where in 2005 the Lukashenko regime launched a campaign against the Polish ethnic minority.[53]

Poland is the sixth most populous member state of the European Union and, ever since joining in 2004, has pursued policies to increase its role in European affairs. Poland has a grand total of 51 representatives in the European Parliament and in addition to this, since 14 July 2009, former Prime Minister of Poland Jerzy Buzek, has been President of the European Parliament.[54]

Administrative divisions

Poland's current voivodeships (provinces) are largely based on the country's historic regions, whereas those of the past two decades (to 1998) had been centred on and named for individual cities. The new units range in area from less than 10000 square kilometres (**unknown operator: u'strong'** sq mi) for Opole Voivodeship to more than 35000 square kilometres (**unknown operator: u'strong'** sq mi) for Masovian Voivodeship. Administrative authority at voivodeship level is shared between a government-appointed voivode (governor), an elected regional assembly (*sejmik*) and an executive elected by that assembly.

The voivodeships are subdivided into *powiats* (often referred to in English as counties), and these are further divided into *gminas* (also known as communes or municipalities). Major cities normally have the status of both *gmina* and *powiat*. Poland currently has 16 voivodeships, 379 powiats (including 65 cities with *powiat* status), and 2,478 *gminas*.

West Pomeranian
Pomeranian
Warmian-Masurian
Podlaskie
Lubusz
Greater Poland
Kuyavian-Pomeranian
Lower Silesian
Opole
Silesian
Świętokrzyskie
Łódź
Masovian
Lublin
Lesser Poland
Subcarpathian

Voivodeship		Capital city or cities
in English	*in Polish*	
Greater Poland	*Wielkopolskie*	Poznań
Kuyavian-Pomeranian	*Kujawsko-Pomorskie*	Bydgoszcz / Toruń
Lesser Poland	*Małopolskie*	Kraków
Łódź	*Łódzkie*	Łódź
Lower Silesian	*Dolnośląskie*	Wrocław
Lublin	*Lubelskie*	Lublin
Lubusz	*Lubuskie*	Gorzów Wielkopolski / Zielona Góra
Masovian	*Mazowieckie*	Warsaw
Opole	*Opolskie*	Opole
Podlaskie	*Podlaskie*	Białystok
Pomeranian	*Pomorskie*	Gdańsk
Silesian	*Śląskie*	Katowice
Subcarpathian	*Podkarpackie*	Rzeszów
Świętokrzyskie (Holy Cross)	*Świętokrzyskie*	Kielce
Warmian-Masurian	*Warmińsko-Mazurskie*	Olsztyn
West Pomeranian	*Zachodniopomorskie*	Szczecin

Military

The Polish armed forces are composed of four branches: Land Forces (*Wojska Lądowe*), Navy (*Marynarka Wojenna*), Air Force (*Siły Powietrzne*) and Special Forces (*Wojska Specjalne*). The military is subordinate to the Minister for National Defence, however its sole commander in chief is the President of the Republic.

Polish Air Force F-16 Fighting Falcon

The Polish army currently consists of 65,000 active personnel, whilst the navy and air force respectively employ 14,300 and 26,126 servicemen and women. The Polish Navy is one of the larger navies on the Baltic Sea and is mostly involved in Baltic operations such as search and rescue provision for the section of the Baltic under Polish command, as well as hydrographic measurements and research; recently however, the Polish Navy played a more international role as part of the 2003 invasion of Iraq, providing logistical support for the United States Navy. The current position of the Polish Air Force is much the same; it has routinely taken part in Baltic Air Policing assignments, but otherwise, with the exception of a number of units serving in Afghanistan, has seen no active combat since the end of the Second World War. In 2003, the F-16C Block 52 was chosen as the new general multi-role fighter for the air force, the first deliveries taking place in November 2006; it is expected (2010) that the Polish Air Force will create three squadrons of F-16s, which will all be fully operational by 2012.

Polish Army vehicles and troops on patrol in Afghanistan

The most important mission of the armed forces is the defence of Polish territorial integrity and Polish interests abroad.[55] Poland's national security goal is to further integrate with NATO and European defence, economic, and political institutions through the modernisation and reorganisation of its military.[55] Currently the armed forces is being re-organised according to NATO standards, and as of 1 January 2010, the transition to an entirely contract-based military has been completed. Previously male citizens were expected to complete a period of active service with the military; since 2007 up until the amendment of the law on conscription, the obligatory term of service was nine months.[56]

Polish military doctrine reflects the same defensive nature as that of its NATO partners. From 1953 to 2009 Poland was a large contributor to various United Nations peacekeeping missions.[55] [57] The Polish Armed Forces took part in the 2003 invasion of Iraq, deploying 2,500 soldiers in the south of that country and commanding the 17-nation Multinational force in Iraq.

The military was temporarily, but severely, affected by the loss of many of its top commanders in the wake the 2010 Polish Air Force Tu-154 crash near Smolensk, Russia, which killed all 96 passengers and crew, including, amongst others, the Chief of the Polish Army's General Staff Franciszek Gągor and Polish Air Force commanding general Andrzej Błasik. They were en route from Warsaw to attend an event to mark the 70th anniversary of the Katyn massacre, whose site is commemorated approximately 19 km west of Smolensk.[58] [59]

Law enforcement and emergency services

Poland has a highly developed system of law enforcement with a long history of effective policing by the State Police Service. The structure of law enforcement agencies within Poland is a multi-tier one, with the State Police providing criminal-investigative services, Municipal Police serving to maintain public order and a number of other specialised agencies, such as the Polish Border Guard, acting to fulfil their assigned missions. In addition to these state services, private security companies are also common, although they possess no powers assigned to state agencies, such as, for example, the power to make an arrest or detain a suspect.

Mounted officers of the State Police in Chorzów

Emergency services in Poland consist of the Emergency Medical Services, Search and Rescue units of the Polish Armed Forces and State Fire Service. Emergency medical services in Poland are, unlike other services, provided for by local and regional government.

Since joining the European Union all of Poland's emergency services have been undergoing major restructuring and have, in the process, acquired large amounts of new equipment and staff.[60] All emergency services personnel are now uniformed and can be easily recognised thanks to a number of innovative design features, such as reflective paint and printing, present throughout their service dress and vehicle liveries. In addition to this, in an effort to comply with EU standards and safety regulations, the police and other agencies have been steadily replacing and modernising their fleets of vehicles; this has left them with thousands of new automobiles, as well as many new aircraft, boats and helicopters.[61]

Economy

Financial centre of Warsaw

Poland's high-income economy[62] is considered to be one of the healthiest of the post-Communist countries and is currently one of the fastest growing within the EU. Having a strong domestic market, low private debt, flexible currency, and not being dependent on a single export sector, Poland is the only European economy to have avoided the late-2000s recession.[63] Since the fall of the communist government, Poland has steadfastly pursued a policy of liberalising the economy and today stands out as a successful example of the transition from a centrally planned economy to a primarily market-based economy. In 2009 Poland had the highest GDP growth in the EU. As of February 2012, the Polish economy has not entered a recession in the wake of the global financial crisis.[64] [65]

The privatization of small and medium state-owned companies and a liberal law on establishing new firms have allowed the development of an aggressive private sector. As a consequence, consumer rights organizations have also appeared. Restructuring and privatisation of "sensitive sectors" such as coal, steel, rail transport and energy has been continuing since 1990. Between 2007 and 2010, the government plans to float twenty public companies on the Warsaw Stock Exchange, including parts of the coal industry. The biggest privatisations have been the sale of the national telecoms firm Telekomunikacja Polska to France Télécom in 2000, and an issue of 30% of the shares in Poland's largest bank, PKO Bank Polski, on the Polish stockmarket in 2004.

Poland is part of the Schengen Area and the EU single market.

The Polish banking sector is the largest in central and eastern Europe as well being as the largest and the most highly developed sector of the country's financial markets. It is regulated by the Polish Financial Supervision Authority. During the transformation to a market-oriented economy, the government privatized some banks, recapitalized the rest and introduced legal reforms that made the sector competitive. This has attracted a significant number of strategic foreign investors. Poland's banking sector has approximately 5 domestic banks, a network of nearly 600 cooperative banks and 18 branches of foreign-owned banks. In addition, foreign investors have controlling stakes in nearly 40 commercial banks, which make up 68% of the banking capital.[66]

Poland has a large number of private farms in its agricultural sector, with the potential to become a leading producer of food in the European Union. Structural reforms in health care, education, the pension system, and state administration have resulted in larger-than-expected fiscal pressures. Warsaw leads Central Europe in foreign investment.[67] GDP growth had been strong and steady from 1993 to 2000 with only a short slowdown from 2001 to 2002.

The economy had growth of 3.7% annually in 2003, a rise from 1.4% annually in 2002. In 2004, GDP growth equaled 5.4%, in 2005 3.3% and in 2006 6.2%.[68] According to Eurostat data, Polish PPS GDP per capita stood at 61% of the EU average in 2009.[69]

Gdynia, situated at Gdańsk Bay on the south coast of the Baltic Sea, is an important seaport of Poland.

Although the Polish economy is currently undergoing economic development, there are many challenges ahead. The most notable task on the horizon is the preparation of the economy (through continuing deep structural reforms) to allow Poland to meet the strict economic criteria for entry into the Eurozone. According to the Polish foreign minister Radosław Sikorski the country could join the eurozone before 2016.[70] Some businesses may already accept the euro as payment. In addition, the ability to establish and conduct business easily has been cause for economic hardship as the World Economic Forum recently ranked Poland near the bottom of OECD countries in terms of the clarity, efficiency and neutrality of its legal framework for firm to settle disputes.[71] A report concluded that on-going foreign business disputes issues may "have damaged Poland's reputation as an attractive location for FDI" by reinforcing the impression of "Poland's substandard reputation for maintaining an efficient and neutral framework to settle business disputes involving multinational foreign investors."[72] Ernst & Young's 2010 European attractiveness survey reported that Poland saw a 52% decrease in FDI job creation and a 42% decrease in number of FDI projects since 2008.[73]

Average salaries in the enterprise sector in December 2010 were 3,848 PLN (1,012 euro or 1,374 US dollars)[74] and growing sharply.[75] Salaries vary between the regions: the median wage in the capital city Warsaw was 4,603 PLN (1,177 euro or 1,680 US dollars) while in Kielce it was only 3,083 PLN (788 euro or 1125 US dollars). Differences in salaries in various districts of Poland is even higher and range from 2,020 PLN (517 euro or 737 US dollars) in Kępno County, which is located in Greater Poland Voivodeship to 5,616 (1,436 euro or 2,050 US dollars) in Lubin County, which lies in Lower Silesian Voivodeship.[76]

According to a Credit Suisse report, Poles are the second wealthiest (after Czechs) of the Central European peoples.[77] [78] This makes Poland an attractive destination for many guest workers particularly from Ukraine, Belarus, Russia and Vietnam.[79] [80] [81] Even though Poland is rather an ethnically homogeneous country, the number of foreigners is growing every year.[81] [82]

Since the United Kingdom, Ireland and some other European countries opened their job markets for Poles, many workers, especially from rural regions, have left the country to seek a better wages abroad. However, there is a rapid growth of the salaries, booming economy, strong value of Polish currency, and quickly decreasing unemployment (from 14.2% in May 2006 to 6.7% in August 2008).[83] Commodities produced in Poland include: electronics, cars (Arrinera, Leopard), buses (Autosan, Solaris, Solbus), helicopters (PZL Świdnik), transport equipment, locomotives, planes (PZL Mielec), ships, military engineering (including tanks, SPAAG systems), medicines (Polpharma, Polfa), food, clothes, glass, pottery (Bolesławiec), chemical products and others.

Corporations

Poland is recognised as a regional economic power within Central Europe, possessing nearly 40 percent of the 500 biggest companies in the region (by revenues).[84] Poland was the only member of the EU to avoid the recession of the late 2000s, a testament to the Polish economy's stability.[63] The country's most competitive firms are components of the WIG20 which is traded on the Warsaw Stock Exchange.

The Warsaw Stock Exchange is, by market capitalisation, one of Central Europe's largest

Well known Polish brands include, amongst others, PKO BP, PKN Orlen, PGE, PZU, PGNiG, Tauron Group, Lotos Group, KGHM Polska Miedź, Telekomunikacja Polska, Plus, Play, PLL LOT, Poczta Polska, PKP, Biedronka, BRE Bank, Getin Holding and TVP.[85]

Poland is recognised as having an economy with significant development potential, overtaking the Netherlands in mid-2010 to become Europe's sixth largest economy.[86] Foreign Direct Investment in Poland has remained strong ever since the country's re-democratisation following the Round Table Agreement in 1989. Despite this, problems do exist, and further progress in achieving success depends largely on the government's privatisation of Poland's remaining state industries and continuing development and modernisation of the economy.

Warsaw is home to many of Poland's largest business enterprises

The list includes the largest companies by turnover in 2009, but does not include major banks or insurance companies:

Rank in 2009[87]	Name of concern	Location of headquarters	Revenue (Thou. PLN)	Profit (Thou. PLN)	Employees
1.	PKN Orlen SA	Płock	49,876,074	1,907,812	4,482
2.	PGE SA	Lublin	22,418,183	5,378,534	46,357
3.	Fiat Auto Poland SA	Bielsko-Biała	19,788,511	820,420	6,421
4.	PGNiG SA	Warsaw	19,493,756	1,442,103	31,685
5.	PBG SA	Poznań	17,761,936	N/A	27,032
6.	Metro Group Poland	Warsaw	16,800,000	N/A	24,077
7.	Telekomunikacja Polska SA	Warsaw	16,560,000	1,597,000	27,667
8.	Lotos Group SA	Gdańsk	15,379,754	1,091,950	4,949
9.	Tauron Group SA	Katowice	13,859,135	1,165,020	28,839
10.	KGHM Polska Miedź SA	Lubin	12,286,107	3,070,799	18,370

Tourism

Poland is a major part of the global tourism market and is currently experiencing an upward trend in its number of visitors; this began shortly after joining the European Union.[88] Tourism in Poland contributes to the country's overall economy and makes up a relatively large proportion of the country's service market.

The city of Zamość is a UNESCO World Heritage Site

The most attractive urban destinations for tourists are:

- Kraków - The former Polish capital, headquarters and the place of coronation of Polish kings.
- Wrocław - One of the oldest cities in Poland with a very checkered history over the centuries included in the many kingdoms and countries. A huge number of monuments, great architecture, lots of parks. European Capital of Culture 2016.
- Warsaw - Present Polish capital
- Gdańsk
- Poznań
- Lublin
- Toruń

Centennial Hall in Wrocław is a UNESCO World Heritage Site

in addition to these the historic site of the Auschwitz German concentration camp near Oświęcim is a noteworthy place of pilgrimage and a now constitutes a major monument to the prevention of war and suffering.

Popular areas of natural beauty include north Poland's Baltic Sea coast, Masurian Lake District and Białowieża Forest, in southern Karkonosze, Table Mountains, Tatra Mountains and Bieszczady Mountains.

Poland's main tourist offerings are thought to be based around city-sightseeing and extra-urban historical monuments, business trips, qualified tourism, agrotourism, mountain hiking and climbing among others.

Poland was the 17th most visited country by foreign tourists in 2008.[89]

Energy

Żerań power station in Warsaw

The electricity generation sector in Poland is still largely fossil-fuel based. Many power plants nationwide use Poland's position as a major European exporter of coal to their advantage by continuing to use coal as the primary raw material in production of their energy; in 2007, hard bituminous coal contributed 48% of energy generation, brown coal and gas 12% each and oil 23%.[90] Currently the three largest Polish coal mining firms (*Weglokoks, Kompania Węglowa and JSW*) extract around 100 million tonnes of coal annually; all three of these companies are key constituents of the Warsaw Stock Exchange's lead economic indexes.

Renewable forms of energy currently only account for a small proportion of Poland's full energy generation capacity.[91] However, the national government has set targets for the development of renewable energy sources in Poland which should see the portion of power produced by renewable resources climb to 7.5% by 2010 and 15% by 2020. This is to be achieved mainly through the construction of wind farms and a number of hydroelectric stations.

Poland is thought to have around 164,800,000,000 m³ of proven natural gas reserves and around 96,380,000 barrels of proven oil reserves. These reserves are currently attended to and exploited by energy supply companies such as PKN Orlen *(the only Polish company listed in the Fortune Global 500)*. However, due to the small amounts of fossil fuels naturally occurring in Poland not being enough to satisfy the full energy consumption needs of the population and thus need to buy from abroad, the country is considered to be a net importer of oil and natural gas.

Transport

Today transport in Poland is provided by means of rail, road, shipping and air travel. Positioned in Central Europe and with an eastern border compromising the largest external border of the Schengen Area with the rest of East-Central Europe, Poland has long been, and remains a key country through which imports to and exports from the European Union pass.

LOT was the first European airline to introduce the Boeing 787 Dreamliner

Since joining the EU in 2004, Poland has invested large amounts of money into the modernisation of its transport networks. The country now has a developing expressways network composed of motorways such as the A1, A2, A4, A18 and express roads such as the S7, S8. In addition to these newly built roads, many local and regional roads are being rebuilt as part of a national programme to rebuild all roads in Poland.[92]

The A4 Motorway near Kraków

Again, with regard to railways, much the same situation is taking place. The Polish authorities have begun a program by which they hope to increase operating speeds across the entire Polish rail network; this is particularly true of a number of national trunk routes which are expected to soon receive new rolling stock capable of speeds over 200 km/h. Finally, there is a plan to introduce high speed rail to Poland from around 2014. The Polish government recently revealed that it intends to connect all major cities to a future high-speed rail network by 2020.[93] Most intercity rail operations in Poland are operated by PKP Intercity whilst regional trains are run by a number of operators, the largest of which is Przewozy Regionalne.

The air and maritime transport markets in Poland are largely well developed. Poland has a number of international airports; the largest of which is Warsaw Chopin Airport, the primary global hub for LOT Polish Airlines, which is the largest airline of Central and Eastern Europe and one of the world's oldest airlines still in operation today. Seaports exist all along Poland's Baltic coast, with most freight operations using either Gdynia or Gdańsk as their base. Passenger ferries link Poland with Scandinavia all year round; these services are provided from Gdańsk by Polferries, Stena Line from Gdynia and Unity Line from the Port of Świnoujście.

Science

In 1925, the chemist Maria Skłodowska-Curie established the first Radium Institute in Poland.[94]

According to Frost & Sullivan's Country Industry Forecast the country is becoming an interesting location for research and development investments.[95] Multinational companies such as: ABB, Delphi, GlaxoSmithKline, Google, Hewlett–Packard, IBM, Intel, LG Electronics, Microsoft, Motorola, Siemens and Samsung have set up research and development centres in Poland.[96] Over 40 research and development centers and 4,500 researchers make Poland the biggest research and development hub in Central and Eastern Europe.[80] [95] Companies chose Poland because of the availability of highly qualified labour force, presence of universities, support of authorities, and the largest market in Central Europe.[95]

Warsaw's Staszic Palace is home to the Polish Academy of Sciences

Today Poland's tertiary education institutions; traditional universities (found in its major cities), as well as technical, medical, and economic institutions, employ around 61,000 researchers and members of staff. There are around 300 research and development institutes, with about 10,000 researchers. In total, there are around 91,000 scientists in Poland today. However, in the 19th and 20th centuries many Polish scientists worked abroad; one of the greatest of these exiles was Maria Skłodowska-Curie, a physicist and chemist who lived much of her life in France. In the first half of the 20th century, Poland was a flourishing centre of mathematics. Outstanding Polish mathematicians formed the Lwów School of Mathematics (with Stefan Banach, Hugo Steinhaus, Stanisław Ulam) and Warsaw School of Mathematics (with Alfred Tarski, Kazimierz Kuratowski, Wacław Sierpiński). The events of World War II pushed many of them into exile. Such was the case of Benoît Mandelbrot, whose family left Poland when he was still a child. An alumnus of the Warsaw School of Mathematics was Antoni Zygmund, one of the shapers of 20th-century mathematical analysis.

According to a KPMG report[97] 80% of Poland's current investors are content with their choice and willing to reinvest. In 2006, Intel decided to double the number of employees in its research and development centre in Gdańsk.[96]

Communications

The Main Municipal Post Office of Bydgoszcz

The share of the telecom sector in the GDP is 4.4% (end of 2000 figure), compared to 2.5% in 1996. Nevertheless, despite high expenditures for telecom infrastructure (the coverage increased from 78 users per 1,000 inhabitants in 1989 to 282 in 2000).

The value of the telecommunication market is zl 38.2bn (2006), and it grew by 12.4% in 2007 PMR.[98] The coverage mobile cellular is over 1000 users per 1000 people (2007). Telephones—mobile cellular: 38.7 million (Onet.pl & GUS Report, 2007), telephones—main lines in use: 12.5 million (Telecom Team Report, 2005).

With regard to internet access, the most popular ADSL services for home users in Poland are Neostrada provided by TPSA, and Net24 provided by Netia. Business users as well as some home users use Internet DSL TP also offered by TPSA. According to Eurostat, OECD and others, Internet access in Poland is amidst the most expensive in Europe. This is mostly caused by the lack of competitiveness. New operators, such as Dialog and GTS Energis are making their own provider lines and offer more attractive and cheaper service. Recently, the Polish Office of Electronical Communication passed a bill forcing the TPSA to rent 51% of their ADSL lines to other ISPs for 60% lower prices. This move will definitely affect the prices of DSL in Poland.

TP S.A. headquarters in Warsaw

The public postal service in Poland is operated by Poczta Polska (The Polish Post). It was created on October 18, 1558, when king Zygmunt August established a permanent postal route from Kraków to Venice (later also to Wilno) in order to manage affairs in Italy that arose after the death of Queen Bona, his mother. Since then the service was dissolved on a number of occasions, most notably during the partitions of Poland. After regaining independence in 1918, the united territory of Poland was in need of a uniform network of communication. Thus, the interwar period saw the rapid development of the postal system as new services were introduced (e.g. money transfers, payment of pensions, delivery of magazines, air mail). Although during national uprisings and in the course of wars communication was provided mainly through field post, which was subject to military authority, postmen always took active part in the fight for independence by secretly delivering parcels and documents, or by providing vital information about the enemy. Many important events in the history of Poland involved the postal service, like the heroic Defence of the Polish Post Office in Gdańsk in 1939 and the participation of the Polish Scouts' Postal Service in the Warsaw Uprising. During the difficult times of the Second World War, the Polish Post in exile would lift up the spirits of compatriots by issuing postage stamps. Nowadays the service is a modern, functioning state-owned company which provides a number of standard and express delivery options, as well as operating the Polish postal home-delivery service. The postal service is currently expanding into the provision of logistical services.

Demographics

Poland, with 38,116,000 inhabitants,[5] has the eighth-largest population in Europe and the sixth-largest in the European Union. It has a population density of 122 inhabitants per square kilometer (328 per square mile).

Long Market in Gdańsk filled with picturesque Dutch style tenements is a favourite meeting place in the Kashubian capital.

Poland historically contained many languages, cultures and religions on its soil. The country had a particularly large Jewish population prior to World War II, when the Nazi Germany's regime led to The Holocaust. It caused Poland's Jewish population, estimated at 3 million before the war, to drop to just 300,000. The outcome of the war, particularly the shift of Poland's borders to the area between the Curzon Line and the Oder-Neisse line, coupled with post-war expulsion of minorities, significantly reduced the country's ethnic diversity. Over 7 million Germans fled or were expelled from the Polish side of the Oder-Neisse boundary.[99]

According to the 2002 census, 36,983,700 people, or 96.74% of the population, consider themselves Polish, while 471,500 (1.23%) declared another nationality, and 774,900 (2.03%) did not declare any nationality. The largest minority nationalities and ethnic groups in Poland are Silesians (173,153 according to the census), Germans (152,897 according to the census, 92% in Opole Voivodeship and Silesian Voivodeship), Belarusians (c. 49,000), Ukrainians (c. 30,000), Lithuanians, Russians, Roma, Jews, Lemkos, Slovaks, Czechs, and Lipka Tatars.[100] Among foreign citizens, the Vietnamese are the largest ethnic group, followed by Greeks and Armenians.

The Main Market Square in Kraków is the heart of Poland's southern cultural capital

The Polish language, part of the West Slavic branch of the Slavic languages, functions as the official language of Poland. Until recent decades Russian was commonly learned as a second language but has been replaced by English and German as the most common second languages studied and spoken.[101]

In recent years, Poland's population has decreased because of an increase in emigration and a sharp drop in the birth rate. Since Poland's accession to the European Union, a significant number of Poles have emigrated, primarily to the United Kingdom, Germany and Ireland in search of work and better work opportunities abroad. In April 2007, the Polish population of the United Kingdom had risen to approximately 300,000, and estimates place the Polish population in Ireland at 65,000. Some sources claim that the number of Polish citizens who emigrated to the UK after 2004 is as high as 2 million.[102] This, however, is contrasted by a recent trend that shows that more Poles are entering the country than leaving it.[103]

Polish minorities are still present in the neighboring countries of Ukraine, Belarus, and Lithuania, as well as in other countries (see Poles for population numbers). Altogether, the number of ethnic Poles living abroad is estimated to be around 20 million.[104] The largest number of Poles outside of Poland can be found in the United States.[105]

Religion

Until World War II Poland and Prussia had a Protestant majority, although a religiously diverse society, in which substantial Jewish, Christian Orthodox and Roman Catholic minorities coexisted. As a result of the Holocaust and the post–World War II flight and expulsion of German and Ukrainian populations, Poland has become overwhelmingly Roman Catholic. In 2007, 88.4% of the population belonged to the Catholic Church.[106] Though rates of religious observance are lower, at 52%[107] or 51% of the Polish Catholics,[108] Poland remains one of the most devoutly religious countries in Europe.[109]

Basilica of Our Lady of Licheń

Holy Spirit Orthodox Church in Białystok

From 16 October 1978 until his death on 2 April 2005 Karol Józef Wojtyła (later Pope John Paul II), a natural born Pole, reigned as Supreme Pontiff of the Roman Catholic Church and Sovereign of Vatican City. His was the second-longest documented pontificate; only Pope Pius IX served longer. He has been the only Slavic and Polish Pope to date, and was the first non-Italian Pope since Dutch Pope Adrian VI in 1522.[110] Additionally he is credited with having played a significant role in hastening the downfall of communism in Poland and throughout Central and Eastern Europe; he is famously quoted as having, at the height of communism in 1979, told Poles "not be afraid", later praying: "Let your Spirit descend and change the image of the land... this land".[111] [112] He is a deeply revered figure within Polish society, and his passing in 2005 was met with large-scale outpourings of national grief.

Religious minorities include Polish Orthodox (about 506,800),[5] various Protestants (about 150,000),[5] Jehovah's Witnesses (126,827),[5] Eastern Catholics, Mariavites, Polish Catholics, Jews, and Muslims (including the Tatars of Białystok). Members of Protestant churches include about 77,500 in the largest Evangelical-Augsburg Church,[5] and a similar number in smaller Pentecostal and Evangelical churches.

Kraków's Tempel Synagogue is one of the largest in Poland still functioning today.

Freedom of religion is now guaranteed by the 1989 statute of the Polish Constitution,[113] enabling the emergence of additional denominations.[114] However, because of pressure from the Polish Episcopate, the exposition of doctrine has entered the public education system as well.[115] [116] According to a 2007 survey, 72% of respondents were not opposed to religious instruction in public schools; alternative courses in ethics are available only in one percent of the entire public educational system.[117]

Famous sites of Christian pilgrimage in Poland include the Monastery of Jasna Góra in the southern Polish city of Częstochowa, as well as the Family home of John Paul II in Wadowice just outside of Kraków.

Health

Poland's healthcare system is based on an all-inclusive insurance system. State subsidised healthcare is available to all Polish citizens who are covered by this general health insurance program. However, it is not compulsory to be treated in a state-run hospital as a number of private medical complexes do exist nationwide.[118]

All medical service providers and hospitals in Poland are subordinate to the Polish Ministry of Health, which provides oversight and scrutiny of general medical practice as well as being responsible for the day to day administration of the healthcare system. In addition to these roles, the ministry is also tasked with the maintenance of standards of hygiene and patient-care.

Main clinical building of the Gdańsk University Medical Centre in Wrzeszcz

Hospitals in Poland are organised according to the regional administrative structure, resultantly most towns have their own hospital *(Szpital Miejski)*. Larger and more specialised medical complexes tend only to be found in major cities, with some even more specialised units located only in the capital, Warsaw. However, all voivodeships have their own general hospital (most have more than one), all of which are obliged to have a trauma centre; these types of hospital, which are able to deal with almost all medical problems are called 'regional hospitals' *(Szpital Wojewódzki)*. The last category of hospital in Poland is that of specialised medical centres, an example of which would be the Skłodowska-Curie Institute of Oncology, Poland's leading, and most highly specialised centre for the research and treatment of cancer.

The Polish health-care industry is currently undergoing a major transformation, with many hospitals being listed as top priorities for refurbishment.[119] As a result of this process, many hospitals have already been thoroughly modernised throughout and are now equipped with the latest in medical hardware. The overall quality of healthcare provision nationwide, as judged by European standards, is generally regarded as being very high.[120] This is reflected in the nation's average life expectancy, which at 71 for males and 80 for females,[121] has shown a marked increase from 63/68 in 2003, and now corresponds with the average figures for life expectancy in the European Union.

Education

The education of Polish society was a goal of rulers as early as the 12th century, and Poland soon became one of the most educated countries in Europe. The library catalogue of the Cathedral Chapter of Kraków dating back to 1110 shows that in the early 12th-century Polish intellectuals had access to European literature. The Jagiellonian University, founded in 1364 by King Casimir III in Kraków, is one of Europe's oldest universities. In 1773 King Stanisław August Poniatowski established the Commission of National Education (*Komisja Edukacji Narodowej*), the world's first state ministry of education.

The *Collegium Maius* is the oldest building of the Jagiellonian University in Kraków

The Adam Mickiewicz University, Poznań

The first university in Poland, Kraków's Jagiellonian University, was established in 1364 by Casimir III the Great in Kraków. It is the oldest university in Poland. It is the second oldest university in Central Europe and one of the oldest universities in the world. Casimir III realized that the nation needed a class of educated people, especially lawyers, who could codify the country's laws and administer the courts and offices. His efforts to found an institution of higher learning in Poland were finally rewarded when Pope Urban V granted him permission to open the University of Krakow.

Since changes made in 2009 education in Poland starts at the age of five or six for the 0 class (Kindergarten) and six or seven years in the 1st class of primary school (Polish *szkoła podstawowa*). It is compulsory that children do one year of formal education before entering 1st class at no later than 7 years of age.

The wearing of traditional academic dress is an important feature of Polish educational ceremonies

At the end of 6th class when the students are 13, they take a compulsory exam that will determine to which lower secondary school (*gimnazjum, pronounced gheem-nah-sium*) (Middle School/Junior High) they will be accepted. They will attend this school for three years for classes, 7, 8, and 9. They then take another compulsory exam to determine the upper secondary level school they will attend. There are several alternatives, the most common being the three years in a *liceum* or four years in a technikum. Both end with a maturity examination (matura, quite similar to French baccalauréat), and may be followed by several forms of upper education, leading to licencjat or inżynier (the Polish Bologna Process first cycle qualification), magister (second cycle qualification) and eventually doktor (third cycle qualification).[122]

The University of Wrocław at night.

There are currently 18 fully accredited traditional universities in Poland, these are then further supplemented by 20 technical universities, nine independent medical universities and five universities for the study of economics. In addition to these institutions there are then nine agricultural academies, three pedagogical universities, a theological academy and three maritime service universities. Poland's long history of promoting the arts has led to the establishment of a number of higher educational institutes dedicated to the teaching of the arts. Amongst these are the seven higher state academies of music. All of these institutions are further supplemented by a large number of private educational institutions and the four national military academies (two for the army and one for each of the other branches of service), bringing the total number of organisations for the pursuit of higher education to well over 500, one of the largest numbers in Europe. The Programme for International Student Assessment, coordinated by the OECD, currently ranks Poland's educational system as the 23rd best in the world, being neither significantly higher nor lower than the OECD average.[123]

Culture

The culture of Poland is closely connected with its intricate 1000 year history[124] Its unique character developed as a result of its geography at the confluence of European cultures. With origins in the culture of the Proto-Slavs, over time Polish culture has been profoundly influenced by its interweaving ties with the Germanic, Latinate and Byzantine worlds as well as in continual dialog with the many other ethnic groups and minorities living in Poland.[125] The people of Poland have traditionally been seen as hospitable to artists from abroad and eager to follow cultural and artistic trends popular in other countries. In the 19th and 20th centuries the Polish focus on cultural advancement often took precedence over political and economic activity. These factors have contributed to the versatile nature of Polish art, with all its complex nuances.[125]

Famous people

Poland is the birthplace of many distinguished personalities (see. List of Polish people), among which are: Mikołaj Kopernik,[126] Fryderyk Chopin,[127] [128] Maria Skłodowska Curie,[129] Stanislaw Ulam, Ernest Malinowski, Tadeusz Kościuszko, Kazimierz Pułaski, Józef Piłsudski, Lech Wałęsa and Pope John Paul II (Karol Wojtyła). Great Polish painter Jan Matejko devoted his monumental art to the most significant historical events on Polish lands, along with the playwright, painter and poet Stanisław Wyspiański. Stanisław Ignacy Witkiewicz (Witkacy) was an example of a Polish avant-garde philosopher and author of aesthetic theories.

Mikołaj Kopernik
(Latin: *Nicolaus Copernicus*)

Polish literature dates back to the 12th century[130] and includes many famous poets and writers such as Jan Kochanowski, Adam Mickiewicz, Bolesław Prus, Juliusz Słowacki, Witold Gombrowicz, Stanisław Lem and, Ryszard Kapuściński. Writers Henryk Sienkiewicz, Władysław Reymont, Czesław Miłosz, Wisława Szymborska have each won the Nobel Prize in Literature. Also a renowned Polish novelist, who wrote in the English language, was Joseph Conrad.[131]

Many world famous Polish movie directors include Academy Awards winners Roman Polański, Andrzej Wajda, Zbigniew Rybczyński, Janusz Kamiński, Krzysztof Kieślowski, Agnieszka Holland. World renowned actresses were Helena Modjeska and Pola Negri.

Poles have outstanding achievements in mountaineering in the Himalayas, especially in the winter gathering Eight-thousander. The most famous Polish climbers are Jerzy Kukuczka, Krzysztof Wielicki, Piotr Pustelnik, Andrzej Zawada, Maciej Berbeka and Wanda Rutkiewicz.

Society

The Grand Hotel in Sopot. Poland's Baltic Sea resorts are very popular tourist destinations amongst Poles as well as foreign tourists

Pope John Paul II is considered to have been a great promoter of Poland around the world

Poland has a great, long standing tradition of tolerance towards minorities, as well as absence of discrimination on the grounds of religion, nationality or race. This is for historical reasons and that is ethnic minorities in Poland are very few. It has a high level of gender equality, promotes disability rights movement and promotes equality. Poland is the first country in the world where corporal punishment was prohibited.[132] Poland has, throughout most of its long history, experienced only very limited immigration from abroad; this trend can largely be attributed to Poland's lack of slavery and overseas colonies as well as its lack of existence as a state during much of the 19th and early 20th centuries. Despite this, the country has for a long time been regarded as having a very tolerant society, which affords equal rights to all people no matter what their ethnic background. This can be said to stem largely from the reign of King Casimir III the Great and his acceptance for Poland's Jewish community, in a time when the most of Europe recessed antisemitic mood. The history of Jews in Poland shows peaceful co-existence of a nation and particular ethnic group.

As many as 96.7% of Polish citizens declare to be Poles, and 97.8% declare that they speak Polish at home (Census 2002). The population of Poland became one of the most ethnically homogeneous in the world as a result of the radically altered borders after World War II and the subsequent migrations. This homogeneity is a result of post World War II deportations ordered by the Soviet authorities, who wished to remove the sizeable Polish minorities from Lithuania, Belarus and Ukraine and repatriation of Ukrainians from Poland to the Soviet Union (see territorial changes of Poland and historical demography of Poland for details). Unlike in many other countries, the ethnic minority rights in Poland are guaranteed directly in the Constitution of Poland (art. 35), and today there are, amongst others, sizeable German, Ukrainian and Belarusian minorities present in the country.[133]

After the formal collapse of Communism in 1989, Poland greatly improved its image in the world and thus has received further support from the country's recent economic success and effective entry into the structures of the European Union. Polish citizens have obtained a good reputation as workers in the united Europe, mainly due to the broad range of jobs beyond the borders of their state, since 2004. The results of an Organization for Security and Co-operation in Europe (OSCE) survey from 2004 showed that Poles work the second most hours per week of any nationality worldwide.

Poland remains one of the most peaceful countries in the world.[134]

Music

Fryderyk Chopin, composer

Artists from Poland, including famous composers like Chopin or Penderecki and traditional, regionalized folk musicians, create a lively and diverse music scene, which even recognizes its own music genres, such as poezja śpiewana and disco polo. As of 2006, Poland is one of the few countries in Europe where rock and hip hop dominate over pop music, while all kinds of alternative music genres are encouraged.

The origins of Polish music can be traced as far back as the 13th century; manuscripts have been found in Stary Sącz, containing polyphonic compositions related to the Parisian Notre Dame School. Other early compositions, such as the melody of *Bogurodzica* and *Bóg się rodzi* (a coronation polonaise for Polish kings by an unknown composer), may also date back to this period, however, the first known notable composer, Mikołaj z Radomia, was born and lived in the 15th century. During the 16th century, two main musical groups – both based in Kraków and belonging to the King and Archbishop of the Wawel – led to the rapid development of Polish music. Composers writing during this period include Wacław z Szamotuł, Mikołaj Zieleński, and Mikołaj Gomółka. Diomedes Cato, a native-born Italian who lived in Kraków from about the age of five, became one of the most famous lutenists at the court of Sigismund III, and not only imported some of the musical styles from southern Europe, but blended them with native folk music.[135]

Tomasz Stańko is a popular contemporary Polish jazz musician

At the end of the 18th century, Polish classical music evolved into national forms like the polonaise. In the 19th century the most popular composers were: Józef Elsner and his pupils Fryderyk Chopin and Ignacy Dobrzyński. Important opera composers of the era were Karol Kurpiński and Stanisław Moniuszko whilst the list of famous soloists and composers included Henryk Wieniawski, Juliusz Zarębski. At the turn of the 19th and 20th centuries the most promiment composers could said to have been Władysław Zeleński and Mieczysław Karłowicz, with Karol Szymanowski gaining prominence prior to World War II. Alexandre Tansman lived in Paris but had strong connections with Poland. Witold Lutosławski, Henryk Górecki, and Krzysztof Penderecki composed in Poland, Andrzej Panufnik emigrated.

Traditional Polish folk music has had a major effect on the works of many well-known Polish composers, and no more so than on Fryderyk Chopin, a widely recognised national hero of the arts. All of Chopin's works involve the piano and are technically demanding, emphasising nuance and expressive depth. As a great composer, Chopin invented the musical form known as the instrumental ballade and made major innovations to the piano sonata, mazurka, waltz, nocturne, polonaise, étude, impromptu and prélude, he was also the composer of a number of polonaises which borrowed heavily from traditional Polish folk music. It is largely thanks to him that the such pieces gained great popularity throughout Europe during the 19th century. Nowadays the most distinctive folk music can be heard in the towns and villages of the mountainous south, particularly in the region surrounding the winter resort town of Zakopane.

Today Poland has a very active music scene, with the jazz and metal genres being particularly popular amongst the contemporary populace. Polish jazz musicians such as Krzysztof Komeda, created a unique style, which was most famous in 1960s and 1970s and continues to be popular to this day. Since the fall of Communism, Poland has become a major venue for large-scale music festivals, chief among which are the Open'er Festival, Opole Festival and Sopot Festival.

Media

Headquarters of TVP in Warsaw

Poland has instituted freedom of press since the fall of communism, a system under which the media was heavily politically controlled and censored. However, public TV and radio are still regulated by the government, this is exercised through an agency called *Krajowa Rada Radiofonii i Telewizji* (*The National Radio and Television Committee*), which is similar to television regulatory commissions in other developed nations.

Poland has a number of major media outlets, chief amongst which are the national television channels. TVP is Poland's public broadcasting corporation; about a third of its income comes from a broadcast receiver licence, while the rest is made through revenue from commercials and sponsorships. State television operates two mainstream channels, TVP 1 and TVP 2, as well as regional programs (TVP Info) for each of the country's 16 voivodeships. In addition to these general channels, TVP runs a number of genre-specific programmes such as TVP Sport, TVP Historia, TVP Kultura and TVP Seriale; there are currently plans to run channels dedicated to the coverage of political affairs (TVP Parlament) and entertainment (TVP Rozrywka).

Studio set of TVP Wrocław's 'Fakty' news program

Poland has a number of internationally broadcast and 24 hour news channels, chief amongst which are Polsat News, TVN 24, and TV Polonia, the latter is a state-run channel dedicated to the transmission of Polish language television for the Polish diaspora abroad. There are a number of major private television outlets such as Polsat and the TVN network.

Poland has a highly developed printed news industry, with daily newspapers like Gazeta Wyborcza *(The Electoral Gazette)*, Rzeczpospolita *(The Republic)* and Gazeta Polska Codziennie providing more traditional, intellectually stimulating reporting and tabloids such as Fakt providing more sensationalist writing which is less current affairs orientated. Rzeczpospolita is one of the nation's oldest publications still in operation today, founded in 1920, it has become a stalwart bastion of Polish reporting and in 2006 won a prestigious award for being, along with the Guardian (a British daily), the best designed newspaper in the world.[136]

Major media outlets are experiencing an ongoing restructuring which is seeing many of them amalgamated into major media groups; a prime example of which is the German Axel Springer AG Publishing conglomerate's purchase of Fakt. International cooperation is also a growing trend within Polish media; TVP recently began cooperating with the French-German TV network ARTE.

Literature

Polish literature has a long and complicated history. During the Middle Ages most Polish authors and academics (Jan Długosz) wrote only in Latin, as at the time, this was the 'academic' language which linked Europe together; Jan Kochanowski broke this trend and became the first author to write the majority of his works in the Polish language. A number of Polish authors have won great renown in the past few centuries, however, this largely stems from the initial success of the works of Adam Mickiewicz, who wrote the first Polish epic, Pan Tadeusz, in 1834. Influential authors of the late 19th and 20th centuries include Henryk Sienkiewicz, Władysław Reymont, Witold Gombrowicz and Czesław Miłosz. To date four Polish authors have won the Nobel Prize in Literature, with Władysław Reymont being one of only nine writers to receive the prestigious award for one particular, outstanding literary work *(awarded for the great national epic, 'The Peasants' in 1924)* rather than their career as a whole.[137]

Adam Mickiewicz (1798–1855)	Henryk Sienkiewicz (1846–1916)	Władysław Reymont (1865–1925)	Witold Gombrowicz (1904–1969)	Czesław Miłosz (1911–2004)

With regard to poetry, Poland has a long and distinguished history of producing world-class poets. Chief among these are the 'three bards' (trzej wieszcze), Mickiewicz, Krasiński and Słowacki; the three national poets of Polish Romantic literature. Incidentally, the Polish word *Wieszcz* means 'prophet' or 'soothsayer', a fitting reference for the three, as the bards were thought to not only voice Polish national sentiments but to frequently foresee the nation's future.

Today the traditions of Polish literature and poetry are being carried forward by a new generation of writers. Included within this group is Wisława Szymborska, a best-selling author and recipient of the 1996 Nobel Prize in Literature.[138]

Architecture

Polish cities and towns reflect the whole spectrum of European styles. Romanesque architecture is represented by St. Andrew's Church in Kraków, and characteristic for Poland Brick Gothic by St. Mary's Church in Gdańsk. Richly decorated attics and arcade loggias are the common elements of the Polish Renaissance architecture,[139] [140] like in City Hall in Poznań. For some time the late renaissance, so called mannerism, most notably in Bishop's Palace in Kielce, coexisted with the early baroque like in Church of SS. Peter and Paul in Kraków.

Renaissance City Hall in Poznań

The Gothic St. Mary's Basilica on the Main Market Square in Kraków

History has not been kind to Poland's architectural monuments. Nonetheless, a number of ancient structures had survived: castles, churches, and stately buildings, often unique in the regional or European context. Some of them have been painstakingly restored, like Wawel Castle, or completely reconstructed after being destroyed in the Second World War, including the Old Town and Royal Castle in Warsaw and the Old Town of Gdańsk. The architecture of Gdańsk is mostly of the Hanseatic variety, a Gothic style common amongst the former trading cities along the Baltic sea and in the northern part of Central Europe. The architectural style of Wrocław is mainly representative of German architecture, since it was for centuries located within the German states. The centre of Kazimierz Dolny on the Vistula is a good example of a well-preserved medieval town. Poland's ancient capital, Kraków, ranks among the best-preserved Gothic and Renaissance urban complexes in Europe. Meanwhile, the legacy of the Kresy Marchlands of Poland's eastern regions, where Wilno and Lwów (now *Vilnius* and *Lviv*) were recognised as two major centres for the arts, played a special role in the development of Polish architecture, with Catholic church architecture deserving special note.[125]

The second half of the 17th century is marked by baroque architecture. Side towers, visible in Branicki Palace in Białystok are typical for Polish baroque. The classical Silesian baroque is represented by the University in Wrocław. Profuse decorations of Branicki Palace in Warsaw are characteristic of rococo style. The centre of Polish classicism was Warsaw under the rule of the last Polish king Stanisław August Poniatowski.[141] The Palace on the Water is the most notable example of Polish neoclassical architecture. Lublin Castle represents the Gothic Revival style in architecture, while the Izrael Poznański Palace in Łódź is an example of eclecticism.

Cuisine

Polish cuisine has influenced the cuisines of its surrounding countries. For centuries the Polish foods have been the arena of competing with France and Italy, due to influx of merchants from Genoa, Florence and other world cities.[142] It is rich in meat, especially pork, chicken and beef, winter vegetables (cabbage in the dish bigos) and spices. Pasta is also featured in many Polish dishes. Most notable are pierogi. Polish national cuisine shares similarities with other European traditions. Generally speaking, Polish cuisine is hearty. The preparation of traditional dishes can be time-intensive and Poles allow themselves a fair amount of time to prepare and enjoy their festive meals, with some feasts, like the Christmas Eve or Easter breakfast, taking a number of days to prepare in their entirety. In addition, most regions of Poland have their own local gastronomic traditions and distinctive flavours.[143]

Polish bigos at the Wawel Castle café

Notable foods in Polish cuisine include: soups – rosół, barszcz, żurek, krupnik, kapuśniak, zupa pomidorowa (*zupa* means soup in Polish), zupa ogórkowa, zupa grzybowa, flaczki (tripe soup); pierogi, kiełbasa, gołąbki, oscypek, kotlet schabowy, kotlet mielony, bigos, various potato dishes, kanapka, zapiekanka, and many more. Traditional Polish desserts include pączki, faworki, gingerbread, babka and others. Fermented dairy include soured milk, buttermilk and kefir; along with savoury side dishes such as gherkin, pickled cucumber and sauerkraut.

Sports

Football (soccer) is the country's most popular sport, with a rich history of international competition.[144] [145] Track and field, basketball, boxing, ski jumping, fencing, handball, ice hockey, swimming, volleyball, and weightlifting are other popular sports. The golden era of football in Poland occurred throughout the 1970s and went on until the early 1980s when the Polish national football team achieved their best results in any FIFA World Cup competitions finishing 3rd place in the 1974 and 1982 editions. The team won a gold medal in football at the 1972 Summer Olympics and also won two silver medals in 1976 and 1992. Poland, along with Ukraine, hosted the UEFA European Football Championship in 2012.[146]

National Stadium, Warsaw, home of Poland national football team and host of Euro 2012

The Polish men's national volleyball team is ranked 4th in the world and the women's volleyball team is ranked 15th. Mariusz Pudzianowski is a highly successful strongman competitor and has won more World's Strongest Man titles than any other competitor in the world, winning the event in 2008 for the fifth time. The first Polish Formula One driver, Robert Kubica, has brought awareness of Formula One Racing to Poland. Poland has made a distinctive mark in motorcycle speedway racing thanks to Tomasz Gollob, a highly successful Polish rider. The national speedway team of Poland is one of the major teams in international speedway and is very successful in various competitions.[147]

The Polish mountains are an ideal venue for hiking, skiing and mountain biking and attract millions of tourists every year from all over the world.[89] Baltic beaches and resorts are popular locations for fishing, canoeing, kayaking and a broad-range of other water-themed sports.

International rankings

The following are links to international rankings of Poland.

Index	Rank	Countries reviewed
Human Development Index 2011	39th	187
Corruption Perceptions Index 2012	41st	176
OECD Working time	2nd	27
Index of Economic Freedom 2012	64th	179
Privacy International Yearly Privacy ranking of countries, 2007	19th	45
Reporters Without Borders Press Freedom Index 2011–2012	24th	179
UNICEF	14th	21
Networked Readiness Index 2010–2011	62nd	138
OICA Automobile Production 2010	18th	48

See also

- Outline of Poland
- Polish migration to the United Kingdom

Notes

a Numerous sources state that Polish Army was the Allies' fourth biggest fighting contingent. Steven J. Zaloga and Richard Hook write that "by the war's end the Polish Army was the fourth largest contingent of the Allied coalition after the armed forces of the Soviet Union, the United States and the United Kingdom."[148] Jerzy Jan Lerski writes "All in all, the Polish units, although divided and controlled by different political orientation, constituted the fourth largest Allied force, after the America, British and Soviet Armies."[149] M. K. Dziewanowski has noted that "if Polish forces fighting in the east and west were added to the resistance fighters, Poland had the fourth largest Allied army in the war (after the USSR, the U.S. and Britain)".[150]

The claim of the fourth biggest Ally needs to be reconsidered, however. Throughout the war, Poland's position varied from the 2nd biggest Ally (after the fall of France, when Polish army outnumbered the French) to perhaps the 5th at the end of it (after the USA, Soviet Union, China and Britain). Please, see the analysis in Polish contribution to World War II.

b Sources vary with regards to what was the largest resistance movement during World War II. The confusion often stems from the fact that as war progressed, some resistance movements grew larger – and other diminished. Polish territories were mostly freed from Nazi German control in the years 1944–1945, eliminating the need for their respective (anti-Nazi) partisan forces (in Poland (although the cursed soldiers continued to fight against the Soviets). Several sources note that Polish Armia Krajowa was the largest resistance movement in Nazi-occupied Europe. Norman Davies wrote: "Armia Krajowa (Home Army), the AK, which could fairly claim to be the largest of European resistance";[151] Gregor Dallas wrote "Home Army (Armia Krajowa or AK) in late 1943 numbered around 400000, making it the largest resistance organization in Europe";[152] Mark Wyman wrote "Armia Krajowa was considered the largest underground resistance unit in wartime Europe".[153] Certainly, Polish resistance was the largest resistance till German invasion of Yugoslavia and invasion of the Soviet Union in 1941. After that point, the numbers of Soviet partisans and Yugoslav partisans begun growing rapidly. The numbers of Soviet partisans quickly caught up and were very similar to that of the Polish resistance.[154] [155] The numbers of Tito's Yugoslav partisans were roughly similar to those of the Polish and Soviet partisans in the first years of the war (1941–1942), but grew rapidly in the latter years, outnumbering the Polish and Soviet partisans by 2:1 or more (estimates give Yugoslavian forces about 800,000 in 1945, to Polish and Soviet forces of 400,000 in 1944).[155] [156]

References

[1] Central Statistical Office – Page 18. "Wyniki Narodowego Spisu Powszechnego Ludności i Mieszkań 2011 Podstawowe informacje o sytuacji demograficzno-społecznej ludności Polski oraz zasobach mieszkaniowych" (http://www.stat.gov.pl/cps/rde/xbcr/gus/PUBL_lu_nps2011_wyniki_nsp2011_22032012.pdf). .

[2] "GUS – Population as of 30.06.2010" (http://www.stat.gov.pl/gus/5840_655_ENG_HTML.htm). Stat.gov.pl. . Retrieved 2011-05-26.

[3] "Poland" (http://www.imf.org/external/pubs/ft/weo/2012/01/weodata/weorept.aspx?pr.x=25&pr.y=3&sy=2009&ey=2012&scsm=1&ssd=1&sort=country&ds=.&br=1&c=964&s=NGDPD,NGDPDPC,PPPGDP,PPPPC,LP&grp=0&a=). International Monetary Fund. . Retrieved 2012-04-20.

[4] "Table 1 – Human Development Index and its components" (http://hdr.undp.org/en/media/HDR_2010_EN_Table1.pdf). *Human Development Index 2010*. Human Development Reports. . Retrieved 4 November 2010.

[5] "Concise Statistical Yearbook of Poland, 2008" (http://www.stat.gov.pl/cps/rde/xbcr/gus/PUBL_maly_rocznik_statystyczny_2008.pdf) (PDF). Central Statistical Office (Poland). 28 July 2008. . Retrieved 2008-08-12.

[6] NationMaster.com 2003–2007, Poland, Facts and figures (http://www.nationmaster.com/country/pl-poland)

[7] Lukowski, Jerzy; Zawaszki, Hubert (2001). *A Concise History of Poland* (First Edition ed.). University of Stirling Libraries – Popular Loan (Q 43.8 LUK): Cambridge University Press. p. 3. ISBN 0-521-55917-0.

[8] "Poland – UNESCO World Heritage Centre" (http://whc.unesco.org/en/statesparties/pl). Whc.unesco.org. . Retrieved 2012-02-06.

[9] (English) "Human Development Index and its components" (http://hdr.undp.org/en/media/HDR_2010_EN_Table1.pdf). *hdr.undp.org*. . Retrieved 2011-08-27.

[10] **Polenia** by Thietmar of Merseburg *Chronicle*, 1002. (German: *Polen*)

[11] **Polani** by John Canaparius, *Vita sancti Adalberti episcopi Pragensis*, or *Life of St. Adalbert of Prague*, 999.

[12] "fr. pal, pele, altd. pal, pael, dn. pael, sw. pale, isl. pall, bre. pal, peul, it. polo, pole, pila, [in:] A dictionary of the Anglo-Saxon languages. Joseph Bosworth. S.275.; planus, plain, flat; from Indo- Germanic pele, flat, to spread, also the root of words like plan, floor, and field. [in:] John Hejduk. Soundings. 1993. p. 399"; "the root pele is the source of the English words "field" and "floor". The root "plak" is the source of the English word "flake" [in:] Loren Edward Meierding. Ace the Verbal on the SAT. 2005. p. 82

[13] Maciej Kosiński, Magdalena Wieczorek-Szmal (2007). "Z mroku dziejów. Kultura Łużycka" (http://www.muzeumczestochowa.pl/wp-content/uploads/2012/11/folder.pdf) (PDF file, direct download 1.95 MB). Muzeum Częstochowskie. Rezerwat archeologiczny (Museum of Częstochowa). pp. 3–4. ISBN 978-83-601281-1-4. . Retrieved January 9, 2013. "... kultura łużycka nie tworzyła jednej zwartej całości. Jak się wydaje, jej skład etniczny był niejednorodny."

[14] Teeple, J. B. (2002). *Timelines of World History*. Publisher: DK Adult.

[15] Jerzy Wyrozumski – *Historia Polski do roku 1505* (History of Poland until 1505), Państwowe Wydawnictwo Naukowe (Polish Scientific Publishers PWN), Warszawa 1986, ISBN 978-83-01-03732-1

[16] (English) Norman Davies (1996). *Europe: a history*. Oxford University Press. p. 428. ISBN 0-19-820171-0. "By 1490 the Jagiellons controlled Poland-Lithuania, Bohemia, and Hungary, but not the Empire."

[17] " Jagiellon dynasty (European history) (http://www.britannica.com/EBchecked/topic/299331/Jagiellon-dynasty)". Encyclopædia Britannica.

[18] Davies (2007). *Warfare, State and Society on the Black Sea Steppe,1500–1700.* (http://books.google.com/books?id=XH4hghHo1qoC&pg=PA17&dq&hl=en#v=onepage&q=&f=false). p.17.

[19] " The Crimean Tatars and their Russian-Captive Slaves (http://www.econ.hit-u.ac.jp/~areastd/mediterranean/mw/pdf/18/10.pdf)" (PDF). Eizo Matsuki, *Mediterranean Studies Group at Hitotsubashi University.*

[20] Józef Andrzej Gierowski – *Historia Polski 1505–1764* (History of Poland 1505–1764), Państwowe Wydawnictwo Naukowe (Polish Scientific Publishers PWN), Warszawa 1986, ISBN 978-83-01-03732-1

[21] Józef Andrzej Gierowski – *Historia Polski 1505–1764* (History of Poland 1505–1764), p. 105-173

[22] " Poland – The 17th-century crisis (http://www.britannica.com/EBchecked/topic/466681/Poland/28190/The-17th-century-crisis)". Britannica Online Encyclopedia.

[23] Józef Andrzej Gierowski – *Historia Polski 1505–1764* (History of Poland 1505–1764), p. 174-301

[24] Józef Andrzej Gierowski – *Historia Polski 1764–1864* (History of Poland 1764–1864), Państwowe Wydawnictwo Naukowe (Polish Scientific Publishers PWN), Warszawa 1986, ISBN 978-83-01-03732-1, p. 1-74

[25] Józef Andrzej Gierowski – *Historia Polski 1764–1864* (History of Poland 1764–1864), p. 74-101

[26] " Poland – Independence Won and Lost, 1914–45. (http://countrystudies.us/poland/13.htm)". Library of Congress Country Studies.

[27] Frątczak, Sławomir Z. (2005). "Cud nad Wisłą" (http://web.archive.org/web/20070708173639/http://www.glos.com.pl/Archiwum_nowe/Rok+2005/032/strona/Cud.html) (in Polish). *Głos* (32/2005). Archived from the original (http://www.glos.com.pl/Archiwum_nowe/Rok+2005/032/strona/Cud.html) on 2007-07-08. . Retrieved June 18, 2006.

[28] Bitter glory: Poland and its fate, 1918 to 1939; p.179

[29] " Russian parliament condemns Stalin for Katyn massacre (http://www.bbc.co.uk/news/world-europe-11845315)". BBC News. November 26, 2010

[30] At the siege of Tobruk

[31] including the capture of the monastery hill at the Battle of Monte Cassino

[32] (English) Lynne Olson & Stanley Cloud. 2003. A Question of Honor. The Kosciuszko Squadron: Forgotten Heroes of World War II. New York: Knopf.

[33] Peszke, Michael Alfred (February 1999). *Poland's Navy, 1918–1945*. Hippocrene Books. p. 37. ISBN 978-0-7818-0672-5.

[34] Stanisław Salmonowicz, *Polskie Państwo Podziemne*, Wydawnictwa Szkolne i Pedagogiczne, Warszawa, 1994, ISBN 978-83-02-05500-3, p.37

[35] The Warsaw Rising (http://www.polandinexile.com/rising.htm), polandinexile.com

[36] Richard J. Kozicki, Piotr Wróbel (eds), *Historical Dictionary of Poland, 966–1945*, Greenwood Press, 1996, ISBN 978-0-313-26007-0, Google Print, p.34 (http://books.google.com/books?id=S6aUBuWPqywC&pg=PA34&ots=VwP7o88St6&dq=Zygmunt+Berling+Warsaw+Uprising&sig=CuHYk98fb5OPXX7ZTyEiFKnuc5I)

[37] Yad Vashem, The Holocaust Martyrs' and Heroes' Remembrance Authority, Righteous Among the Nations – per Country & Ethnic Origin January 1, 2009. Statistics (http://www1.yadvashem.org/righteous_new/statistics.html)

[38] Richard C. Lukas, *Out of the Inferno: Poles Remember the Holocaust*, University Press of Kentucky 1989 – 201 pages. Page 13; also in Richard C. Lukas, *The Forgotten Holocaust: The Poles Under German Occupation, 1939–1944*, University Press of Kentucky, 1986, Google Print, p.13 (http://books.google.ca/books?id=lz9obsxmuW4C&pg=PA13&dq="&sig=ACfU3U0SGgyvqSbL4bypepYoO_CbYc_N_w).

[39] " European Refugee Movements After World War Two (http://www.bbc.co.uk/history/worldwars/wwtwo/refugees_01.shtml)". BBC – History.

[40] (English) "Merkel honours Polish freedom struggle and Tiananmen victims" (http://www.thelocal.de/politics/20090604-19717.html). *www.thelocal.de*. AFP. . Retrieved 2009-09-28.

[41] Arthur Bliss Lane *I saw Poland betrayed: An American Ambassador Reports to the American People*. Indianapolis: The Bobbs-Merrill Company, 1948.

[42] (Polish) Polska. Historia (http://encyklopedia.pwn.pl/haslo.php?id=4575043) PWN Encyklopedia. Retrieved 11 July 2005.

[43] "Real GDP growth in CEECs" (http://transitioneconomies.blogspot.com/2006/05/real-gdp-growth-in-ceecs.html). Transitioneconomies.blogspot.com. 2006-05-28. . Retrieved 2009-05-06.

[44] "WHY POLAND?" (http://polisci.lsa.umich.edu/documents/jjackson/chapt1.pdf) (PDF). . Retrieved 2009-07-08.

[45] "Europe's border-free zone expands" (http://news.bbc.co.uk/1/hi/7153490.stm). BBC News. 21 December 2007. . Retrieved 28 July 2011.

[46] Dyrekcja Generalna Lasów Państwowych, Raport o stanie lasów w Polsce (Report on the Status of Forests in Poland), (http://bip.lasy.gov.pl/pl/bip/px_~raport_o_stanie_lasow_2011.pdf?page_opener=http://bip.lasy.gov.pl/pl/bip/raporty_i_prognozy) *Centrum Informacyjne Lasów Państwowych*. (Polish)

[47] "Poland.pl – White Stork – About White Stork" (http://storks.poland.pl/about_stork/index.htm). Storks.poland.pl. . Retrieved 2009-05-06.

[48] (Polish) "Wybrzeże Morza Bałtyckiego" (http://zalewszczecinski.net/akweny/wybrzeze-morza-baltyckiego). *www.zalewszczecinski.net*. . Retrieved 2009-11-16.

[49] "News Poland | Minorities make the news" (http://www.newpolandexpress.pl/polish_news_story-3634-minorities_make_the_news_.php). Newpolandexpress.pl. 2011-10-15. . Retrieved 2011-11-02.

[50] "Transgender woman poised for seat in Poland's new parliament" (http://www.telegraph.co.uk/news/worldnews/europe/poland/8817279/Transgender-woman-poised-for-seat-in-Polands-new-parliament.html). London: Telegraph. 2011-10-10. . Retrieved 2011-11-02.

[51] Davies, Norman (1996). *Europe: A History* (http://books.google.com/books?id=jrVW9W9eiYMC&pg=PA699). Oxford University Press. p. 699. ISBN 0-19-820171-0. .

[52] Grushenko, Kateryna (12 November 2010). "Polish representative: 'Poland is ready to help Ukraine as long as you are interested'" (http://www.kyivpost.com/news/guide/world-in-uktaine/detail/89674/). Kyiv Post. . Retrieved 28 July 2011.

[53] "Bordering on madness: Belarus mistreats its Polish minority" (http://economist.com/displaystory.cfm?story_id=4085710). The Economist. 16 June 2005. . Retrieved 28 July 2011.

[54] "Jerzy Buzek elected President of the European Parliament" (http://www.europarl.europa.eu/sides/getDoc.do?type=IM-PRESS&reference=20090713IPR58063&language=EN). European Parlament. 2009-07-14. . Retrieved 2010-12-12.

[55] (Polish) "Strategia Bezpieczeństwa Narodowego RP" (http://www.wp.mil.pl/pliki/File/zalaczniki_do_stron/SBN_RP.pdf) (PDF). *www.wp.mil.pl*. . Retrieved 2008-09-26.

[56] Day, Matthew (5 August 2008). "Poland ends army conscription" (http://www.telegraph.co.uk/news/worldnews/europe/poland/2505447/Poland-ends-army-conscription.html). London: Telegraph. . Retrieved 28 July 2011.

[57] "Polska zakończyła udział w misjach po auspicjami ONZ – Wiadomości z kraju i ze świata – Gazeta Prawna – Partner pracodawcy, narzędzie specjalisty" (http://www.gazetaprawna.pl/wiadomosci/artykuly/382706,polska_zakonczyla_udzial_w_misjach_po_auspicjami_onz.html). Gazetaprawna.pl. 2009-12-31. . Retrieved 2011-11-02.

[58] "Accident Database" (http://www.airdisaster.com/cgi-bin/view_details.cgi?date=04102010&reg=101&airline=Polish+Air+Force). AirDisaster.com. . Retrieved 2010-12-12.

[59] "Senior Polish figures killed in plane crash" (http://news.bbc.co.uk/2/hi/europe/8613395.stm). BBC. 2010-04-11. .

[60] "– 15 tys. zimowych mundurów trafi do jednostek" (http://www.policja.pl/portal/pol/1/59960/15_tys_zimowych_mundurow_trafi_do_jednostek.html). Policja.pl. 30 September 2009. . Retrieved 28 July 2011.

[61] "Nowe radiowozy dla policji" (http://moto.onet.pl/1545847,1,nowe-radiowozy-dla-policji,artykul.html?node=2). Moto.onet.pl. 3 March 2009. . Retrieved 28 July 2011.

[62] "Country and Lending Groups | Data" (http://data.worldbank.org/about/country-classifications/country-and-lending-groups#High_income). Data.worldbank.org. . Retrieved 2010-11-09.

[63] "How Poland became only EU nation to avoid recession" (http://articles.cnn.com/2010-06-29/world/poland.economy.recession_1_poland-transition-economies-eastern-europe?_s=PM:WORLD). CNN. 29 June 2010. . Retrieved 28 July 2011.

[64] "Central Europe Risks Downgrades on Worsening Finances (Update1)" (http://www.bloomberg.com/apps/news?pid=newsarchive&sid=apz_BEuHrNpI). Bloomberg.com. 2009-09-21. . Retrieved 2010-01-27.

[65] "Zloty to Gain, Says LBBW, Most Accurate Forecaster (Update1)" (http://www.bloomberg.com/apps/news?pid=newsarchive&sid=a7k1g_PkuROs#). Bloomberg.com. 2009-10-09. . Retrieved 2010-01-27.

[66] "Banking Sector in Poland | Thomas White International" (http://www.thomaswhite.com/explore-the-world/emerging-market-spotlight/poland-banking.aspx). Thomaswhite.com. 2011-09-16. .

[67] " Poland in the Lead (http://www.warsawvoice.pl/archiwum.phtml/5473/)", *The Warsaw Voice*, September 2002. Retrieved on August 11, 2007.

[68] "Polish Information and Foreign Investment Agency. News" (http://www.paiz.gov.pl/nowosci/?id_news=1297&lang_id=12). *www.paiz.gov.pl*. . Retrieved 2008-09-26.

[69] "GDP per capita in PPS" (http://epp.eurostat.ec.europa.eu/tgm/table.do?tab=table&init=1&plugin=1&language=en&pcode=tsieb010). Eurostat. . Retrieved 2010-06-25.

[70] Bloomberg Businessweek. 2 December 2011. Official: Poland to be ready for euro in 4 years (http://www.businessweek.com/ap/financialnews/D9RCGVPG2.htm)

[71] Schwab, Klaus. "The Global Competitiveness Report 2010-2011" (http://www3.weforum.org/docs/WEF_GlobalCompetitivenessReport_2010-11.pdf). World Economic Forum. . Retrieved 25 April 2011.
[72] Shapiro, Robert J.. "Foreign Direct Investments in Developing Nations: Issues in Telecommunications and the Modernization of Poland" (http://www.investmentsecurity.org/wp-content/uploads/2011/02/CEIS-FDI-Report_Apr-2011.pdf). CEIS. . Retrieved 27 April 2011.
[73] "Waking up to the new economy: Ernst & Young's 2010 European attractiveness survey" (http://www.ey.com/Publication/vwLUAssets/Attractiveness_survey_2010_EU/$FILE/Attractiveness_survey_2010_EU.pdf). Ernst & Young. . Retrieved 25 April 2011.
[74] "Communication on the average monthly salary in enterprise sector excluding payments from profit awards in December 2010" (http://finanse.wp.pl/kat,58434,title,GUS-podal-dane-o-wynagrodzeniach,wid,13051878,wiadomosc.html?ticaid=1ba11) (in **(Polish)**). Stat.gov.pl. . Retrieved 2010-11-04.
[75] "OECD Economic Outlook No. 82 – Poland" (http://web.archive.org/web/20070328163217/http://www.oecd.org/dataoecd/6/32/20213254.pdf) (PDF). Archived from the original (http://www.oecd.org/dataoecd/6/32/20213254.pdf) on 2007-03-28. . Retrieved 2010-04-12.
[76] "Statistic Office of Poland(GUS)" (http://www.stat.gov.pl/bdr_n/app/wybrane_cechy.wymiary) (in **(Polish)**). Stat.gov.pl. . Retrieved 2010-11-19.
[77] "Poles getting rich quick..." (http://www.thenews.pl/business/artykul141422_poles-getting-rich-quick-.html). Polskie Radio, thenews.pl. . Retrieved 2010-10-13.
[78] Jędrzej Bielecki. "Polacy są w światowej czołówce bogacących się narodów" (http://forsal.pl/artykuly/458053,polacy_sa_w_swiatowej_czolowce_bogacacych_sie_narodow.html). Dziennik Gazeta Prawna. . Retrieved 2010-10-13.
[79] http://www.cespi.it/WPMIG/Country%20mig-POLAND.pdf
[80] "Topic Galleries" (http://www.chicagotribune.com/news/sns-rt-us-poland-immigrantstre7bs11w-20111229,0,2982543.story). chicagotribune.com. . Retrieved 2012-02-06.
[81] Kasat Sp. z o.o. (2008-12-31). "Imigranci w Polsce 2008 – eGospodarka.pl – Raporty i prognozy" (http://www.egospodarka.pl/36792,Imigranci-w-Polsce-2008,1,39,1.html). eGospodarka.pl. . Retrieved 2012-02-06.
[82] "Coraz więcej Azjatów emigruje do Polski" (http://fakty.interia.pl/prasa/news/coraz-wiecej-azjatow-emigruje-do-polski,1326394,16). Interia.pl. . Retrieved 2010-10-13.
[83] "Eurostat February 2008 – Euro area unemployment stable at 7.1%" (http://web.archive.org/web/20080726044250/http://epp.eurostat.ec.europa.eu/pls/portal/docs/PAGE/PGP_PRD_CAT_PREREL/PGE_CAT_PREREL_YEAR_2008/PGE_CAT_PREREL_YEAR_2008_MONTH_04/3-01042008-EN-AP.PDF). Web.archive.org. 26 July 2008. Archived from the original (http://epp.eurostat.ec.europa.eu/pls/portal/docs/PAGE/PGP_PRD_CAT_PREREL/PGE_CAT_PREREL_YEAR_2008/PGE_CAT_PREREL_YEAR_2008_MONTH_04/3-01042008-EN-AP.PDF) on 2008-07-26. . Retrieved 28 July 2011.
[84] "Polish economy seen as stable and competitive" (http://www.wbj.pl/article-51029-polish-economy-seen-as-stable-and-competitive.html). Warsaw Business Journal. 9 September 2010. . Retrieved 28 July 2011.
[85] ATLAS AMONGST THE BEST POLISH BRANDS (http://dziennik.dziennikbudowy.pl/ang.nsf/0/2CE31E39A26A64F3C1256C31002C1FA4?OpenDocument). Retrieved 21 November 2010.
[86] "fDi: Poland Primed for Golden Decade" (http://www.ginannebrownell.com/2010/10/fdi-poland-primed-for-golden-decade/). GinanneBrownell.com. 8 October 2010. . Retrieved 28 July 2011.
[87] **(Polish)** "Lista 500 największych polskich firm" (http://www.lista500.polityka.pl/rankings/show). *www.lista500.polityka.pl*. . Retrieved 2011-08-27.
[88] **(English)** "Travel And Tourism in Poland" (http://www.euromonitor.com/Travel_And_Tourism_in_Poland). *www.euromonitor.com*. . Retrieved 2009-10-12.
[89] **(English)** "UNTWO World Tourism Barometer, Vol.5 No.2" (http://www.tourismroi.com/Content_Attachments/27670/File_633513750035785076.pdf). *www.tourismroi.com*. . Retrieved 2009-10-12.
[90] "Use of renewable fuel in Central and Eastern Europe" (http://www.frost.com/prod/servlet/cpo/205103205.htm). Frost.com. 1 March 2009. . Retrieved 28 July 2011.
[91] "EU Commission – Energy factsheet P74" (http://ec.europa.eu/energy/publications/statistics/doc/2010_energy_transport_figures.pdf) (PDF). . Retrieved 2011-07-28.
[92] "National Road Rebuilding Program (Polish)" (http://bip.mswia.gov.pl/portal/bip/175/17721/Narodowy_Program_Przebudowy_Drog.html). Bip.mswia.gov.pl. 2006-02-16. . Retrieved 2011-07-28.
[93] "Super pociągi zamiast autostrad (Polish)" (http://www.tvn24.pl/12690,1634983,0,1,super-pociagi-zamiast-autostrad,wiadomosc.html). *TVN24*. December 23, 2009. . Retrieved December 25, 2009.
[94] **(English)** Richard Francis Mould (1993). *A century of X-rays and radioactivity in medicine: with emphasis on photographic records of the early years* (http://books.google.com/?id=IXPz7bVR7g0C&printsec=frontcover&dq=A+century+of+x-rays+and+radioactivity+in+medicine:&q=). p. 19. ISBN 978-0-7503-0224-1. .
[95] Newswire Poland Emerges as the European R&D Hub Despite Favorable Conditions in Asia Pacific (http://www.newswiretoday.com/news/26490/)
[96] Polish Information and Foreign Investment Agency Poland – R&D centre (http://web.archive.org/web/20080219145055/http://www.paiz.gov.pl/index/?id=7b7a53e239400a13bd6be6c91c4f6c4e)
[97] **(English)** KPMG Sp. z o.o.. "Why Poland?" (http://www.paiz.gov.pl/files/?id_plik=7513). *www.paiz.gov.pl*. p. 3. . Retrieved 2011-08-27. "Over 80% of foreign investors see the results of their investments to date as positive or very positive and none of the studied

companies reported a negative opinion."

[98] (English) "Key data on IT and telecoms market in Poland, 2004–2006" (http://web.archive.org/web/20061108212651/http://www.sat.org.au/reviews/articles_pl_middle_ages.htm). *www.itandtelecompoland.com*. Archived from the original (http://www.itandtelecompoland.com/index.php?item=2) on 2006-11-08. . Retrieved 2008-09-24.

[99] *The Expulsion of 'German' Communities from Eastern Europe at the end of the Second World War* (http://cadmus.iue.it/dspace/bitstream/1814/2599/1/HEC04-01.pdf), Steffen Prauser and Arfon Rees, European University Institute, Florense. HEC No. 2004/1. p.29

[100] (English) Michał Buchowski, Katarzyna Chlewińska. "Tolerance and Cultural Diversity Discourses in Poland" (http://www.eui.eu/Projects/ACCEPT/Documents/Research/wp1/ACCEPTPLURALISMWp1BackgroundreportPoland.pdf). *www.eui.eu*. . Retrieved 2011-08-27.

[101] (English) Jan Repa (2007-01-05). "Poles return to Russian language" (http://news.bbc.co.uk/2/hi/europe/6233821.stm). *news.bbc.co.uk*. . Retrieved 2011-08-27. "In former satellite countries like Hungary or Poland, knowledge of Russian dwindled rapidly – to be replaced by English and German."

[102] Doughty, Steve (2006-04-25). "UK lets in more Poles than there are in Warsaw" (http://www.dailymail.co.uk/news/article-384121/UK-lets-Poles-Warsaw.html). London: Dailymail.co.uk. . Retrieved 2010-04-12.

[103] (English) Alexi Mostrous, Christine Seib (February 16, 2008). "Tide turns as Poles end great migration" (http://www.timesonline.co.uk/tol/news/uk/article3378877.ece). *www.timesonline.co.uk* (London). . Retrieved 2011-08-27. "The Times has established that, for the first time since they began arriving en masse four years ago, more UK-based Poles are returning to their homeland than are entering Britain."

[104] "Polish Diaspora (Polonia) Worldwide" (http://culture.polishsite.us/articles/art79fr.htm). Culture.polishsite.us. . Retrieved 2010-04-12.

[105] "Centers of Polish Immigration in the World – USA and Germany" (http://culture.polishsite.us/articles/art90fr.htm). Culture.polishsite.us. 2003-03-15. . Retrieved 2010-04-12.

[106] "Maly Rocznik Statystyczny Polski 2009" (http://www.stat.gov.pl/cps/rde/xbcr/gus/PUBL_oz_maly_rocznik_statystyczny_2009.pdf) (in (Polish)) (PDF). . Retrieved 2009-09-26.

[107] "*94% Polaków wierzy w Boga*" (http://www.ekumenizm.pl/content/article/20080925183042429.htm). Ekumenizm.pl. 2008-09-25. . Retrieved 2010-04-12.

[108] Weekly Mass Attendance of Catholics in Nations with Large Catholic Populations, 1980–2008 (http://web.archive.org/web/20080214110918/http://cara.georgetown.edu/bulletin/international.htm) – World Values Survey (WVS)

[109] (Polish) Centrum Badania Opinii Społecznej (*Centre for Public Opinion Research (Poland)* CBOS). Komunikat z badań; Warszawa, Marzec 2005. Co łączy Polaków z parafią? (http://www.cbos.pl/SPISKOM.POL/2005/K_049_05.PDF) Preface. Retrieved 2007-12-14.

[110] Wilde, Robert. "Pope John Paul II 1920–2005" (http://europeanhistory.about.com/od/religionandthought/a/biojohnpaulii.htm). About.com. . Retrieved 2009-01-01.

[111] Domínguez, Juan: 2005

[112] "Pope John Paul II and Communism" (http://www.religion-cults.com/pope/communism.htm). Public domain text. May be distributed freely. No rights reserved.. . Retrieved 2009-01-01.

[113] (Polish) Dr Zbigniew Pasek, Jagiellonian University, "Wyznania religijne" (http://web.archive.org/web/20061128165649/http://www.religioznawstwo.uj.edu.pl/syllabusy/pasek-wrwp.rtf). Archived from the original (http://www.religioznawstwo.uj.edu.pl/syllabusy/pasek-wrwp.rtf) on November 28, 2006. . Retrieved 2007-09-15. Further reading: Ustawa o gwarancjach wolności sumienia i wyznania z dnia 17 V 1989 z najnowszymi nowelizacjami z 1997 roku.

[114] (Polish) Michał Tymiński, "Kościół Zielonoświątkowy" (http://web.archive.org/web/20050102151031/http://www.kz.pl/index.php?p=13&id=3&i=8). Archived from the original (http://www.kz.pl/index.php?p=13&id=3&i=8) on January 2, 2005. . Retrieved 2007-09-14.

[115] (Polish) Dr. Paweł Borecki, "Opinia prawna dotycząca religii w szkole" (http://www.racjonalista.pl/kk.php/s,5534). Kateda Prawa Wyznaniowego Uniwersytetu Warszawskiego. . Retrieved 2007-09-14.

[116] (Polish) Wirtualna Polska, Wiadomości. "Polacy przeciwni wliczaniu ocen z religii do średniej" (http://wiadomosci.wp.pl/kat,9911,wid,9125933,wiadomosc.html?ticaid=1478d). . Retrieved 2007-09-14.

[117] (Polish) Olga Szpunar, "Dorośli chcą religii w szkole" (http://miasta.gazeta.pl/krakow/1,35798,4360977.html). Gazeta Wyborcza Kraków. . Retrieved 2007-09-15.

[118] "Poland Guide: The Polish health care system, An introduction: Poland's health care is based on a general" (http://www.justlanded.com/english/Poland/Poland-Guide/Health/The-Polish-health-care-system). Justlanded.com. . Retrieved 2011-07-28.

[119] "Polish hospitals" (http://polandpoland.com/polish_hospitals.html). Polandpoland.com. . Retrieved 2011-07-28.

[120] "The Quality of Medical Treatment and Surgery in Poland" (http://www.articlesbase.com/medicine-articles/the-quality-of-medical-treatment-and-surgery-in-poland-602240.html). Articlesbase.com. . Retrieved 2011-07-28.

[121] "WHO | Poland" (http://www.who.int/countries/pol/en/). Who.int. 2011-05-17. . Retrieved 2011-07-28.

[122] OECD (2009). "The impact of the 1999 education reform in Poland" (http://www.pisa.oecd.org/dataoecd/50/26/45721631.doc). . Retrieved 2010-09-17.

[123] (English) "Range of rank on PISA 2006 science scale" (http://www.oecd.org/dataoecd/42/8/39700724.pdf) (PDF). *www.oecd.org*. . Retrieved 2008-09-24.

[124] Adam Zamoyski, The Polish Way: A Thousand Year History of the Poles and Their Culture (http://www.amazon.co.uk/Polish-Way-Thousand-History-Culture/dp/978-0-7818-0200-0). Published 1993, Hippocrene Books, Poland, ISBN 978-0-7818-0200-0

[125] Ministry of Foreign Affairs of Poland, 2002–2007, AN OVERVIEW OF POLISH CULTURE. (http://www.poland.gov.pl/Culture,484.html) Access date 12-13-2007.

[126] (English) "Nicolaus Copernicus" (http://www.britannica.com/EBchecked/topic/136591/Nicolaus-Copernicus). *www.britannica.com*. . Retrieved 2008-10-10.

[127] (French) Rey Alain (1993). *Le petit Robert 2 : (dictionnaire universel des noms propres, alphabétique et analogique)*. INIST-CNRS, Cote INIST : L 22712: Le Robert, Paris, FRANCE. ISBN 978-2-85036-210-1.

[128] (English) Michael Kennedy, ed. (2004). *The Concise Oxford dictionary of music*. Oxford University Press. ISBN 978-0-19-860884-4. p. 141

[129] (French) "Maria Sklodowska. La jeunesse" (http://mariecurie.science.gouv.fr/portrait/portrait1_1.php). *mariecurie.science.gouv.fr*. . Retrieved 2008-10-10.

[130] (Polish) Koca, B. (2006). "Polish Literature – The Middle Ages (Religious writings)" (http://web.archive.org/web/20061108212651/http://www.sat.org.au/reviews/articles_pl_middle_ages.htm). Archived from the original (http://www.sat.org.au/reviews/articles_pl_middle_ages.htm) on 2006-11-08. . Retrieved 10 December 2006.

[131] (English) Zdzislaw Najder (1998). "Profiles – Joseph Conrad" (http://www.culture.pl/en/culture/artykuly/os_conrad_joseph). *www.culture.pl*. . Retrieved 2008-09-30.

[132] http://www.coe.int/t/dg3/children/corporalpunishment/pdf/EnglishQuestionAnswer_en.pdf

[133] "Dr. Sławomir Łodziński, Helsinki Foundation for Human Rights, "The Protection of National Minorities in Poland"" (http://www.minelres.lv/reports/poland/poland_NGO.htm). Minelres.lv. . Retrieved 2011-07-28.

[134] http://www.visionofhumanity.org/wp-content/uploads/2011/10/2011GPIMethodologyResultsFindings.pdf

[135] (English) "The Music Courts of the Polish Vasas" (http://www.semper.pl/muzyczne_dwory_summary.pdf). *www.semper.pl*. p. 244. . Retrieved 2009-05-13.

[136] Busfield, Steve (2006-02-21). "Guardian wins design award" (http://media.guardian.co.uk/site/story/0,,1714643,00.html). *The Guardian* (London). .

[137] "Facts on the Nobel Prize in Literature" (http://nobelprize.org/nobel_prizes/literature/shortfacts.html). Nobelprize.org. 2009-10-05. . Retrieved 2011-07-28.

[138] Adam Gopnik (June 5, 2007). "Szymborska's 'View': Small Truths Sharply Etched" (http://www.npr.org/templates/story/story.php?storyId=10721773). npr.org. . Retrieved December 12, 2010.

[139] (English) "Szydłowiec" (http://web.archive.org/web/20060623020259/http://www.szydlowiec.pl/grafika/index/szydl1.pdf). *www.szydlowiec.pl*. p. 9. Archived from the original (http://www.szydlowiec.pl/grafika/index/szydl1.pdf) on June 23, 2006. . Retrieved 2009-04-23.

[140] Many designs imitated the arcaded courtyard and arched loggias of the Wawel palace. (English) Michael J. Mikoś. "RENAISSANCE CULTURAL BACKGROUND" (http://www.staropolska.pl/ang/renaissance/Mikos_renaissance/Cultural_r.html). *www.staropolska.pl*. p. 9. . Retrieved 2009-04-23.

[141] (English) John Stanley (March–June 2004). "Literary Activities and Attitudes in the Stanislavian Age in Poland (1764–1795): A Social System?" (http://findarticles.com/p/articles/mi_qa3763/is_200403/ai_n9363971/?tag=content;col1). *findarticles.com*. . Retrieved 2009-04-23.

[142] Jerzy Pasikowski (2011). "Wpływy kuchni innych narodów na kształt kuchni polskiej (Foreign influences in Polish cuisine)" (http://newsgastro.pl/jerzy-pasikowski-radzi/88908-wpywy-kuchni-innych-narodow-na-ksztat-kuchni-polskiej.html). Portal Gastronomiczny *NewsGastro*. . Retrieved January 7, 2013.

[143] Encyklopedia kuchni: "Kuchnia polska." (http://www.kuchnia.tv/kuchnia_encyklopedia-kuchni_185_kuchnia-polska.html) Kuchnia.tv, ul. W. Sikorskiego 902-758 Warszawa, 2011.

[144] "FIFA World Cup Statistics-Poland" (http://www.fifa.com/worldfootball/statisticsandrecords/associations/association=pol/worldcup/index.html). FIFA. . Retrieved December 12, 2010.

[145] "FIFA Statistics – Poland" (http://www.fifa.com/worldfootball/statisticsandrecords/associations/association=pol/othertournaments/index.html). . Retrieved December 12, 2010.

[146] "Poland hosts Euro 2012!" (http://www.warsaw-life.com/poland/euro-2012). warsaw-life.com. . Retrieved December 12, 2010.

[147] "Speedway World Cup: Poland win 2010 Speedway World Cup" (http://www.worldspeedway.com/artman/publish/article_13423.shtml). worldspeedway.com. . Retrieved December 18, 2010.

[148] Steven J. Zaloga; Richard Hook (21 January 1982). *The Polish Army 1939–45* (http://books.google.com/books?id=AAdYFeW2fnoC&pg=PA3). Osprey Publishing. pp. 3–. ISBN 978-0-85045-417-8. . Retrieved 6 March 2011.

[149] Jerzy Jan Lerski (1996). *Historical dictionary of Poland, 966–1945* (http://books.google.com/books?id=QTUTqE2difgC&pg=PA18). Greenwood Publishing Group. pp. 18–. ISBN 978-0-313-26007-0. . Retrieved 6 March 2011.

[150] E. Garrison Walters (1988). *The other Europe: Eastern Europe to 1945* (http://books.google.com/books?id=64VpSBd7xUcC&pg=PA276). Syracuse University Press. pp. 276–. ISBN 978-0-8156-2440-0. . Retrieved 6 March 2011.

[151] Norman Davies, *God's Playground: A History of Poland*, Columbia University Press, 2005, ISBN 0-231-12819-3, Google Print p.344 (http://books.google.com/books?id=EBpghdZeIwAC&pg=PA344&dq="Armia+Krajowa"+largest&ei=hTrMR_W-G4mWzASVs9GtCQ&sig=iE7xbtRu3rvEsVZZgCeUsqEqj6s)

[152] Gregor Dallas, *1945: The War That Never Ended*, Yale University Press, 2005, ISBN 0-300-10980-6, Google Print, p.79 (http://books.google.com/books?id=LXdVF6LmTa8C&pg=PA79&dq="Armia+Krajowa"+largest&as_brr=3&ei=RjvMR6KnPJPAzAT-ppWvCQ&

sig=Ksba8pTs5pu55YiAqseCLy6Kl5k)
[153] Mark Wyman, *DPs: Europe's Displaced Persons, 1945–1951*, Cornell University Press, 1998, ISBN 0-8014-8542-8, Google Print, p.34 (http://books.google.com/books?id=lHNw7MnsmlYC&pg=PA34&dq="Armia+Krajowa"+largest&lr=&as_brr=3&ei=NzzMR_mOIJGSzQSb7cSwCQ&sig=kv3oN5z3YgAgcT8Vgy4aIFRHknE)
[154] See for example: Leonid D. Grenkevich in The Soviet Partisan Movement, 1941–44: A Critical Historiographical Analysis, p.229 or Walter Laqueur in The Guerilla Reader: A Historical Anthology, (New York, Charles Scribiner, 1990, p.233.
[155] Velimir Vukšić (23 July 2003). *Tito's partisans 1941–45* (http://books.google.com/books?id=SLix5hc4WRgC&pg=PA11). Osprey Publishing. pp. 11–. ISBN 978-1-84176-675-1. . Retrieved 1 March 2011.
[156] Anna M. Cienciala, THE COMING OF THE WAR AND EASTERN EUROPE IN WORLD WAR II. (http://web.ku.edu/~eceurope/hist557/lect16.htm), History 557 Lecture Notes

External links

Movie (on-line)

- Animated history of Poland, (PARP, Expo 2010 Shanghai China) (http://www.youtube.com/watch?v=2DrXgj1NwN8)
- Borders of Poland, AD. 990–2008 (http://www.youtube.com/watch?v=eAVVWlUywO0)
- Poland under German occupation 1939–1945 (http://www.youtube.com/watch?v=nSzejCWXKhA)

General

- Poland.gov.en – Polish national portal (http://en.poland.gov.pl/)
- Polish Information (http://www.paiz.gov.pl/index/?lang_id=1)
- Ministry of Foreign Affairs (http://www.msz.gov.pl/)
- Chief of State and Cabinet Members (https://www.cia.gov/library/publications/world-leaders-1/world-leaders-p/poland.html)
- Poland (https://www.cia.gov/library/publications/the-world-factbook/geos/pl.html) entry at *The World Factbook*
- Poland (http://ucblibraries.colorado.edu/govpubs/for/poland.htm) at *UCB Libraries GovPubs*
- History of Ancient Poland (http://www.theancientweb.com/explore/content.aspx?content_id=24)
- Poland (http://www.dmoz.org/Regional/Europe/Poland/) at the Open Directory Project
- Poland profile (http://www.bbc.co.uk/news/world-europe-17753718) from the BBC News
- Poland (http://www.oecd.org/poland/) OECD
- Poland (http://europa.eu/about-eu/countries/member-countries/poland/index_en.htm) EU
- Poland (http://www.britannica.com/EBchecked/topic/466681/Poland) *Encyclopædia Britannica* entry
- Wikimedia Atlas of Poland
- Geographic data related to Poland (http://www.openstreetmap.org/browse/relation/49715) at OpenStreetMap
- Key Development Forecasts for Poland (http://www.ifs.du.edu/ifs/frm_CountryProfile.aspx?Country=PL) from International Futures

Culture

- Commonwealth of Diverse Cultures: Poland's Heritage (http://www.commonwealth.pl/)

Travel

- Poland's Official Travel Website (http://www.poland.travel/en)
- About Poland (http://www.international-travel-tours.com/poland/about-poland.html)
- Photos from Poland (http://www.klub-beskid.com/album-pologne/index.php?lang=english)

Article Sources and Contributors

Dąb,_Greater_Poland_Voivodeship *Source*: http://en.wikipedia.org/w/index.php?title=D%C4%85b%2C_Greater_Poland_Voivodeship *Contributors*: Kotniski, Stepheng3

Village *Source*: http://en.wikipedia.org/w/index.php?title=Village *Contributors*: 1223Sallybride, 21655, A.Hassen, AKeen, Aarohan Mettu, Abductive, Acsian88, AdamDeanHall, Adamhauner, Adityamadhav83, Airjatt, Aitias, Akrabbim, Al Ameer son, Alaexis, Ale jrb, AlefZet, Aleksandr Grigoryev, Alensha, Altenmann, Amaury, Amberrock, Aminullah, Andrew M. Vachin, Andrewpmk, Andycjp, Animum, Apalamarchuk, Apokrif, Arpingstone, Arthena, AussieLegend, Azreey, Bejinhan, Beland, Belirac, Berig, Big Bird, Bill37212, Biruitorul, Bjh21, Bkonrad, Bldonne, Bobrayner, BocoROTH, Bogdan, Bogdangiusca, Bomac, Bossmanham13, Bossmanham98, Branka France, BreakingFree40, Brian the Editor, Brian0918, Buistr, Butko, C.Fred, Caiaffa, CalJW, CalumMcA, Canley, Capricorn42, Cassowary, Catgut, Catslisten, Cecropia, Cenarium, Ceriy, Cgoodwin, Chasnor15, Chentchen, Chirag, Chris j wood, Christopher.e.dunn, Chzz, Ciudad jardin, Cntras, Colonies Chris, CommonsDelinker, Conleyb15, Conorbrady.ie, Corregere, Cosmic Latte, CrimsonBlue, Cwolfsheep, D6, DITWIN GRIM, Da-drumer, DailyWikiHelp, Daniel Case, Daniel J. Leivick, Daniel Quinlan, David Biddulph, David Kernow, Dbg1233, Ddstretch, DePiep, Deathawk, Deeptrivia, Den fjättrade ankan, Deqon, Deror avi, Desiphral, Dezaree, Dieter Simon, Dinesh.jangra1, Discospinster, Djr13, Doctor Whom, Donarreiskoffer, Doseiai2, Dottod, Dpkipling, Durgesh.Rai, Dysepsion, E ponti, E0steven, Egard89, Ehrenkater, Elassint, Eleassar, Ellywa, Elockid, Elroygoh02, Emyr93, Epipelagic, Erianna, Eu.stefan, EugeneZelenko, EurekaLott, Extra999, Ezhiki, Fang 23, Farosdaughter, Fat pig73, Fayenatic london, Flewis, Francis Tyers, Francs2000, Fred Bauder, Frosted14, Fujurcitook, Funandtrvl, FunkyFly, Futurebird, Gardar Rurak, Garhowell, Gegnome, Gfoley4, Gingermint, Ginkgo100, Giro720, Glenn, Globalphilosophy, GoOdCoNtEnT, Gogo Dodo, Gorai, Grafikm fr, GrahamSmithe, Gtswiki, Guanaco, Guss2, Gustavo Szwedowski de Korwin, Gwen-chan, Gz33, Hahnchen, Halsteadk, Harisu1988, Hayabusa future, Headtoadman, Hebrides, Helenmwarwick, HenryLi, HighSpeed-X, Hmains, Hohum, Hroðulf, Hu12, Hydrogen Iodide, IGeMiNix, IShadowed, IW.HG, Ianisbored, Indiaforever, Ineffable3000, Inwind, Isam, IstvanWolf, IvanKlinko, J.delanoy, JHunterJ, JPSheridan, Jahangard, James086, Janetleopold, Jared Preston, Jcuk, JdeJ, Jeremy Bolwell, Jerrch, Jim1138, Jno, Jongleur100, Julesd, Julia W, JuneGloom07, KR3ColinB, Kajasudhakarababu, Kamiekam, Keelanrush, Kelisi, Kelovy, Khalid Mahmood, Khanvez, Khazar2, Kikos, Kimberry352, Kironbd07, KirrVlad, Kjramesh, Kman543210, Kmsasidhar, Komalrajiana, Ktr101, Kurruption, Kwamikagami, Larucio, Laurinavicius, Ldonna, Leonig Mig, Levineps, Lew121, LiDaobing, Lightmouse, LilHelpa, Linaduliban, Little muddy funkster, Littlealien182, Llywrch, Lonewolf BC, Lord Geznikor, Lozleader, Luk, Lulu1127, Lumos3, Lynda Finn, MacTire02, Maglame, Maitch, Majorly, Malick78, Mallakottai, Mani1, Manly 2008, Marek69, Mariano12 1989, Mariwan zanko, Mark91, Markmark12, Materialscientist, Mayooranathan, Mdzafri, Meldor, Menacheryarun, Mfo7787, MickMacNee, Mike Rosoft, MisfitToys, MisterVodka, Modulatum, Moebiusuibeom-en, Moonriddengirl, Mrmuk, Ms. Hayden and Ms. S, Mtaylor848, Mvjs, Mwalcoff, Mxn, NE2, Naturenet, Naveen Sankar, NerdyScienceDude, Nethergreen55, NewEnglandYankee, Nickshanks, Nistra, Noclevername, Northdeigo, Nricardo, Nyttend, O1ive, Old Moonraker, Orangemike, OscarKosy, P.Marlow, Pankajkumar.9009, Parkwells, Patrick, PaulBannister, Pawyilee, Penrithguy, Peter James, PeterCScott, Pgan002, Phagopsych, Pharaoh of the Wizards, PhilKnight, Pinethicket, PleaseStand, Pokrajac, Ppeace, Pug1776, Puneet 348, Quaysys, R'n'B, RDF, Racerx11, Raeky, Randhirreddy, RandomAct, Rcingham, Realteacher, Reeses14, Reinyday, Rich Farmbrough, Rickyrab, Risker, Rjm at sleepers, Robdav69, Rocket71048576, Rwxrwxrwx, Saccerzd, Sadrettin, Saga City, Same7same7, SamuraiClinton, Samyo, Sansonic, Sardanaphalus, Schinhofen, Schzmo, ScottSteiner, Seb az86556, Seivadnadroj, Sem10efg, Sephiroth BCR, Sfu, Shakingjoker117, Shamsurabiu, Shantavira, Sheddybey, Shell Kinney, Shergo korali, Shoeofdeath, SilkTork, Sjakkalle, Sk jmp, Sketchmoose, Sljaxon, Sole Soul, Speciate, StephenDawson, Steven Zhang, Stevo D, StewartNetAddict, Stifle, Student BSMU, Subhrat, Suffusion of Yellow, Suprgye, Surya Prakash.S.A., Szambo, THATSBETTER, Taamu, Tabletop, TaintedMustard, Tangerinehistry, Tbhotch, Teinesavaii, TejasDiscipulus2, TenIslands, Tenth Plague, TerryAga, The Thing That Should Not Be, The Transhumanist, TheArguer, Theranos, This, that and the other, Thomasrflagel, Tide rolls, Tjanke, Tobby72, Tobias Conradi, Tpbradbury, Triona, Tsuchiya Hikaru, TurkChan, Twanner921, Twin Bird, Tytyty456, Ukabia, UnicornTapestry, Untifler, Utcursch, V2k, VMS Mosaic, Vadim Kiev, Vegaswikian, Vhdwikiuser, Vikramdidawat, Vivek Rai, Vmenkov, WOSlinker, Walex03, Warofdreams, Wayland, Wayne Slam, Wendy1991, West Brom 4ever, Whytecypress, Widr, Wiki Wikardo, Wikijens, William Allen Simpson, William Avery, Winhunter, Wknight94, WojPob, Woohookitty, WorldWide Update, Wrelwser43, Xiaoyu of Yuxi, Xyz or die, Yakudza, Yonkie, Yupik, Zacharie Grossen, Zloyvolsheb, Zoeyzazu, Zollerriia, Zotel, Александър, პაატა შ, , , , 475 anonymous edits

Gmina_Krzymów *Source*: http://en.wikipedia.org/w/index.php?title=Gmina_Krzym%C3%B3w *Contributors*: Andrea Coppola, D6

Adamów,_Gmina_Krzymów *Source*: http://en.wikipedia.org/w/index.php?title=Adam%C3%B3w%2C_Gmina_Krzym%C3%B3w *Contributors*: Kotniski, Stepheng3, Ulric1313

Borowo,_Konin_County *Source*: http://en.wikipedia.org/w/index.php?title=Borowo%2C_Konin_County *Contributors*: Arpingstone, Ulric1313

Konin_County *Source*: http://en.wikipedia.org/w/index.php?title=Konin_County *Contributors*: A bit iffy, Balcer, Darwinek, Elonka, PRB, 5 anonymous edits

Greater_Poland_Voivodeship *Source*: http://en.wikipedia.org/w/index.php?title=Greater_Poland_Voivodeship *Contributors*: A bit iffy, Ajh1492, Al Silonov, Amcde, Appleseed, Balcer, Barticus88, Betacommand, Biruitorul, Bongoramsey, Bonás, Bushy moustache, Bwood, Caius2ga, Chanheigeorge, CharlotteWebb, Chris G, Colonies Chris, CommonsDelinker, Cryptic, David Parker, Dominus Vobisdu, El., Elapsed, Elonka, Emax, Flatterworld, Gnesener1900, Greenshed, GvmBG, Halibutt, Hallihallo61, Joy, Julo, Kintetsubuffalo, Kotniski, Kpjas, LilHelpa, LittleDan, Logologist, Lysy, MJCdetroit, MaGioZal, MacTire02, Marek69, Mazur, Mix321, NaveedLLB, Nihil novi, Odder, Ohconfucius, Orioane, PerV, Piotrus, Plastikspork, PolishPoliticians, ProhibitOnions, R'n'B, Radomil, Sardanaphalus, Scipius, Slawojarek, Socialservice, Spacejam2, Sun Creator, Sylwia Ufnalska, Szczecin, Template namespace initialisation script, Temuri rajavi, Tim!, Topory, Umb is gay, WLKP, Wik, William Allen Simpson, Windharp, Youandme, Zbihniew, Zoe, Јованвб, 43 anonymous edits

Voivodeships_of_Poland *Source*: http://en.wikipedia.org/w/index.php?title=Voivodeships_of_Poland *Contributors*: Abhijitsathe, Adz, Ajh1492, Akerans, Alan Liefting, Alpha Quadrant, Altenmann, Amiraram, Andres, Aotearoa, Appleseed, B7628, Barticus88, BionicWilliam, Biruitorul, Bjung, Bonás, Brisvegas, Brockert, BrownHairedGirl, Byrial, CZmarlin, Caius2ga, Cautious, Chanheigeorge, Christianna1219, Colonies Chris, CrniBombarder!!!, Crusoe8181, DO'Neil, Danim, David Kernow, David Parker, DeirYassin, Deville, Docu, Domino theory, Dominus Vobisdu, Electionworld, Elonka, Emaus, Emax, Erwin, Fayenatic london, Gauss, Gnesener1900, Halibutt, Hmains, JDavid, JNW, Jarekt, Jauhienij, Jopxton, Juzeris, Kelisi, Kotniski, Kpalion, Kpjas, Kwamikagami, LinasLit, Logologist, Lysy, MacTire02, Magioladitis, Marcin Suwalczan, Marek69, Mathiasrex, Maxí, MeltBanana, Meursault2004, Miranche, Naive cynic, Nihil novi, NoychoH, Nuno Tavares, Olessi, Orioane, OwenBlacker, Pak21, Patrick, Phil Boswell, Piotr Osada, Piotrus, Polish29, Poznaniak, Rarelibra, Reaper Eternal, RedWolf, Rjwilmsi, Roke, RollingBarrels, Sangjinhwa, Sanya, Sardanaphalus, Scipius, Severo, Sky Harbor, Skäpperöd, Slawojarek, Spacejam2, Steff, Sterlingsilver850, SwPawel, Sylwia Ufnalska, Tar-ba-gan, Template namespace initialisation script, Topbanana, Ttwaring, Uncle Milty, Valentinian, Vanjagenije, Volunteer Marek, WLKP, Wik, Wmahan, Worobiew, Xx236, Zinneke, 62 anonymous edits

Poland *Source*: http://en.wikipedia.org/w/index.php?title=Poland *Contributors*: 130.94.122.xxx, 217.99.96.xxx, 2eet2eet, 334a, 3sen, 4 bity muzyki, 4darek, 9cds, A bit iffy, A-giau, A. (pl), A.Jalil, A2Kafir, A8UDI, ABCD, ARC Gritt, Aaron Brenneman, Abductive, Abeg92, Academic Challenger, Achaemenes II, Acroterion, Adam Rock, AdamJacobMuller, AdamantlyMike, Adambro, Adamcisme123, Adashiel, Adavis444, Adz, Ae-a, Afinati, Agnus, Agquarx, Aherunar, Ahoerstemeier, AirdishStraus, AirlineMan333, Aivazovsky, Ajh1492, Akamad, Akanemoto, Alawadhi3000, Albanman, Alcatraz014, AlefZet, Aleph-4, Aleph4, Alex.muller, Alex1100, Alexander Domanda, Alexandria, Ali@gwc.org.uk, AliveFreeHappy, AllenZh, Allstarecho, Almankib, Altenmann, Amakuru, Amish 01, Ammon86, Amoruso, Ancheta Wis, Andre Engels, Andy M. Wang, Andy Marchbanks, Andy4789, Andycjp, Andypl, AngelOfSadness, Angela, Anger22, Angr, Annarapo, Annonnimus, Anonymous editor, Anonymous from the 21st century, Another disinterested reader, Antakaa, Antandrus, Antique Rose, Antman, Aotearoa, AppleJordan, Appleseed, Aquintero82, Architect JDF, Arcillaroja, Argentina 678, Ari x, ArielGold, Aries1975, Arjun01, Army1987, Armydepot, Art LaPella, Arthena, Arvis21, Arwel Parry, AshLOVESgnr, Asidemes, Astian, Ataltane, Atlant, Atomekk, Attilios, Atwardow, Audirs8, Auntof6, Ausir, Austrian, Avala, Avaya1, Avb, Avicennasis, AxG, Axegod12, Axt, AzaToth, Azalero, Azar66, AzseicsoK, B-2StealthEditor, BGManofID, BLT000, BRG, Babajobu, Backspace, Balcer, Bardhylius, Baristarim, Barry Kent, Barryob, Bart133, Barticus88, Bartsimpson, Baseball Watcher, Bastian17, Battlecry, Bayerischermann, Bazonka, Bazza 7, Bbatsell, Bbulzibar, Beagel, Beamathan, Beaumont, Beefyt, Beegees, Beetstra, Bejnar, Beland, Ben-Velvel, Bencherlite, Benqish, Benwildeboer, Berkunt, BernardZ, BernardaAlba, Betacommand, Bftravel, Biala Gwiazda, Bigboydog1, Bility, Bill3000, Bjung, Bkell, Black Falcon, Blair P. Houghton, Blood Red Sandman, BlueMars, Blueshade, Bmicomp, BoBo1488, Bob marley 118, Bobblewik, Bobbybrown07, Bobo192, Bobtbuilder, Bocianski, Bodhi141, Boeing720, Bogdan, Bolo1729, Boraczek, Borgarde, Borisblue, Bornkaren, BostonMA, Boud, Boydenmb, Bprsolt Qaoddz, Br77rino, Brabblebrex, Brandmeister (old), Brandon, Briaboru, Brian0918, Brion VIBBER, Brisvegas, BritishWatcher, Brudasek, Brunopioss, Brutannica, Buaidh, Bukubku, Bull Market, BurgererSF, Burschenschafter, Bushy moustache, Bwv1004, C filev, C.Fred, CHJL, CIreland, CLW, CT Cooper, CTF83!, CWY2190, Cactus.man, Caerwine, Caius2ga, Caknuck, CalJW, CalicoCatLover, Caltas, CambridgeBayWeather, Can't sleep, clown will eat me, CanadianPride, CanisRufus, Canterbury Tail, Cantus, Caponer, CaptainFugu, Carbonix, Carlwev, CarolGray, Casliber, Caspian, Cassini83, Catherine-Jayne, Cautious, Celinakat, Celtmist, Century0, CesarB, ChKa, Chabuk, Challisrussia, CharlesMartel, Charlietemps, CharlotteWebb, Chase me ladies, I'm the Cavalry, Cheesefondue, Cheeseloop, ChemTerm, Chensiyuan, Chevinki, Chris G, Chris cufc, Chris the speller, ChrisGriswold, Chrisch, Chrischrischrischris, Chrishmt0423, Chrislk02, Christian List, Christoph.mamok, Christopher Kraus, Chriswilkins2, Chromaticity, Chudzik321, Chumchum7, Chunky Rice, CieloEstrellado, Cimmerian praetor, Civil Engineer III, Closedmouth, CloutierFan02, Cocoaguy, Colchicum, Colonel Mustard, CommonsDelinker, Conscious, Consequencefree, Constantine Gorov, Constantzeanu, Conversion script, Coolszach2, Cormic, Corpx, Corvettegrandsport, Count Iblis, Coveritus, Covervisit, Cptnono, Craftypants, Crazycomputers, Crystallina, Cs-wolves, Csiew88, Cubus1984, Cunningham, Curps, Cusio, Cybergoth, Cybershore, Czas Na Zywiec, Czrisher, Czyrko, D, D6, DBaba, DJ Clayworth, DJPeugeot, DMJohnston, DO'Neil, DRoninLA, DSuser, DVD R W, DX-Kevo, Daddy1234, Daffy123, Damian Radu, Damiang1997, Danaeb, Daniel De Mol, Daniel Quinlan, Daniil naumoff, Danim, Danlock2, Danny, Dannycas, Danziger100, Dariusz Szwed, Darklord517, Darth Kalwejt, Darwinek, Dave.Dunford, Davenbelle, Davewild, David Kernow, David Liuzzo, David Parker, David Richtman, Davidkazuhiro, Dawidbernard, Dbachmann, Dbenbenn, DeadEyeArrow, Deanna Bardenfleth, Deathmetal Greg, Debresser, Deflective, Dekisugi, Delldot, Demiurge, Denkealsobin, Denniss, Der Golem, DerHexer, Derek Ross; DerekJeter, Deschain97, Desiphral, Deuar, Digitalme, Digwuren, Dillard421, Dimkoa, Dina, Diogenes, Director958, Discospinster, Distal24, DivineIntervention, Dj3500, Dmanning, Dockingman, Docta247, Docu, Dodo von den Bergen, Dohaeng, Don Alessandro, Doopdoop, Doprendek, Dougy109, Downyourpants, DrSlony, Dragoman1, Draktorn, Dreadstar, Dru of Id, Drwhorules, DubaiTerminator, Dudeimoncoke, Dust Filter, DuttyYo, Duxwing, Dyamantese, Dycedarg, Dysepsion, Dysprosia, E Pluribus Anthony, EMT1871, ERcheck, ESkog, EagerToddler39, EarthPerson, Eclecticology, Edderso, EddyVadim, Editor at Large, Edivorce, Edward saint-ivan, Edward321, Ehheh, Ejoty, Ejército Rojo 1950, Ek8, Ekuprusevicius, El C, El-042, Eldin, Electionworld, ElectricValkyrie, Elementjunky, Elipongo, Elliskev, Elmondo21st, Elonka, Emax, Emigrant85, Emil Kastberg, Emote, Enchanter, EncycloPetey, Encyclopedist, Enforcer Of Lulz, Entryshare, Enviroboy, Enzo Aquarius, Eob, Ericamick, Erik1980, Esperant, Espoo, Etams, Eu.stefan, EugeneZelenko, Eumolpo, EuroLine, Ev, Evercat, Everyking, Everytime,

Evlekis, FF2010, Fabhcún, Fabian Hassler, Farlieghhungerford, Farosdaughter, Farquaadhnchmn, Fasach Nua, FashionNugget, FayssalF, Fbeauregard, Fbifriday, Femto, Filemon, Filipdr, Filpaul, Filthdafugout, Filyo79, FinnishDriver, FireGaamer1122, Fireaxe888, Firstorm, FlagUploader, Flamarande, Flatterworld, Flockmeal, Fnagaton, Fordan, Fordie123, Foregone conclusion, Formeruser-81, Formulax, Foundert, Fox Mccloud, Fram, Francvs, Freak in the bunnysuit, Freakyfridge, Frees, FreplySpang, Freyr, Frietjes, Frstr, Frymaster, FunPika, Funandtrvl, Funnyhat, Future Perfect at Sunrise, Fvincent, G. Campbell, Gabbe, Gadfium, Gagabrain, Gaius Cornelius, Galati, Galileo01, Gam0001, Garion96, Garraisgreat, Garzo, Gaz, Gazilion, Gazza1133, Gdo01, Geeks123, Geeoharee, Gene Nygaard, Geneb1955, Generalboss3, Geniu, Geopolio, George The Dragon, Georgeslegloupier, Gggh, Gherald, Ghost321, GhostPirate, GhostWhoVotes, Gibble, Gilliam, Gimmetrow, GinaDana, Giraffedata, Giving is good, as long as you're getting, Gnesener1900, Gobonobo, Godfinger, Gogo Dodo, GoingBatty, Golbez, Golem Unity, Goltz, Good Olfactory, Goodlittleangel, GorillazFanAdam, GraemeL, Graham87, Gran2, Grayshi, Green Giant, GregorB, Grendelkhan, Greyhood, Ground Zero, Grsd, Grubber, Gryffindor, Grzechooo, Gscshoyru, Gsklee, Gtg204y, Guinness man, Gurch, Gustavo Szwedowski de Korwin, Gwernol, Gznorneplatz, Gzornenplatz, H, H3xx, H8erade, HIDECCHI001, HJ Mitchell, Hadal, Hagerman, Halibutt, Happyme22, Hardman1, Harmattan, Hatch68, Hayden120, Hdkhdk, Hdt83, Hebereke, Heimstern, Hello32020, Henry Flower, Hephaestos, Hillel, HistoriaPolska, Historian932, HistoricalAce, Hmains, Holod, Homunculus 2, Hoschimobi, Hottentot, Hu12, Husond, Hut 8.5, Ian Pitchford, Ianhowlett, Iapain wiki, Icehawk1223, Icollins68, Icseaturtles, Idaltu, Ief, Ignacy nowopolski, Ignatzmice, Ikiroid, Iksisy, Illythr, Imagod777, Imlepid, InShaneee, Indisciplined, Indon, Infogengachi, Informationguy, Insanely Beautiful, InsanityJ, Intangible2.0, Inter, Interfector, Internet Ignoramus, Iohannes Animosus, Irek Biernat, Irishguy, Istros, Istvan, IstvanWolf, Itai, Italiano111, Ivan89, J.delanoy, J04n, JLaTondre, JRWalko, Jacek Kendysz, Jacoplane, Jacurek, Jadog9495, Jagoperson, Jahanas, JamesBWatson, Jameswilson, JanaMitz, Jankaspar, Janmariamazurczak, Janneman, Janus Shadowsong, JanusTehPheo, Japsai, Jaranda, Jarble, Jared Preston, Jaro7788, Jauerback, Jcaragonv, Jcc7386, Jcrook1987, JdeJ, Jecar, Jeff G., Jeff8765, Jeffhoy, Jeffmaylortx, Jensboot, Jerem43, Jh51681, Jhendin, Jiang, Jihg, Jiminy Krikkitt, Jimmyvanthach, Jirka.h23, Jizg, Jjbaruel, Jkelly, Jklin, Jkw0010, Jlujan69, Jni, JoeBlogsDord, Joffeloff, John, John K, John Price, John Smith M.D., Ph.D, John Vandenberg, John of Reading, Johnkenyon, Joho104, Jon513, Jonas Taylor, Jor, Jorunn, Joseph Solis in Australia, Josephschutz, Josiah Rowe, Jossi, Joy, Jpeob, Jpgordon, Jprg1966, Jrooksjr, Jseidel, Jsport, Jtbelliott, Jukes85, Julekmen, Julo, Jumbuck, Jura2102, Juraj cisarik, Jurand, Jusjih, Jwissick, Jxraynor, KConWiki, KJohansson, KOP, Kaihsu, Kajzderski, Kamil m, Kantokano, KapilTagore, Karasek, Karenjc, Karl, Karol Langner, Kaszkawal, Katarzyna, Kbrooks, Keegan, Keelm, KeithB, Kelisi, Kelly Martin, Kemiv, Khazar, Khoikhoi, King Bigos, Kinneyboy90, Kintetsubuffalo, Kirk diamond, Klimbert, Kman665, Kmcdm, Know Your NME, KnowledgeOfSelf, Knowledgebycoop, Knut, Knverma, Kobold 86, Kochamanita, KokkaShinto, Kollie1, KoloszRodos, Konkursor, Konrii, Konstable, Kopcik, Kotasik, Kotjze, Kotniski, Koyaanis Qatsi, Kozuch, Kpalion, Kpjas, KrakatoaKatie, Krautsaredummies, Krecik001, Kresen, Krich, Krupo, Krystiandl, Krzysiu, Krzyzowiec, Ksenon, Kuba1000, Kubek15, Kubigula, Kubster, Kukini, Kungfuadam, Kuracpm1, Kuru, Kusma, Kwamikagami, Kwpolska, Kyokpae, L95slovenia, LA2, LAX, LMB, LOL, LOTkid, LUCPOL, Laflin22, Laforgue, Lambrook, Lampak, Langus-TxT, Latka, Laudaka, Lcamtuf, Le Fou, Leafyplant, Lear 21, Lebanonman19, Lectonar, LeeHunter, Lemonade100, Lenin and McCarthy, LeonWhite, Levineps, Leyo, Liam Plested, LibLord, LifeStar, Liftarn, Lightdarkness, Lightmouse, Lights, Likeminas, Ling.Nut, Listowy, Llewelyn MT, Llort, LlywelynII, Lockesdonkey, Logan, Logologist, Lolster, Lomisss, LonelyMarble, Looxix, Lord Voldemort, Loren36, Lost tourist, LoveToWiki, Lovesthespooge, Lrschum, Lst27, Luapnampahc, Luca Italy, Lucasm, Ludvikus, LugPaj, Luna Santin, Lycurgus, Lynntoniolondon, Lysy, M.C., MAXXX-309, MD, MER-C, MJCdetroit, MJM74, MMich, MSGJ, MaLu, MackSalmon, MadGeographer, Madriversnowboarder, Maelnuneb, Magister Mathematicae, Mailer diablo, Majorly, Maju wiki, Makemi, Makgraf, MakmoudHassan, Malcolm Farmer, Malcomosx, Malick78, Malik Shabazz, Malo, Mamaberry11, Man vyi, Maor X, Marcin Suwalczan, Marcin.sobczyk, Marco polo, Marek69, MarekF, Mareklug, Mark, MarkGallagher, Markotek, Martarius, Marthainky, Martim33, Martinwilke1980, Marx01, Marzoregulus@hotmail.com, Marzoregulus@yahoo.co.ph, Master of Puppets, Mathiasrex, Matman from Lublin, Matt 989, Mattbr, Matthead, Mauricio Maluff, Mav, Maxamegalon2000, Maxumum12, MałyG, Mb nl, Mc6415, Mc6809e, McDogm, Mcshadypl, Mdann52, Meekywiki, Mega Prime, Meika, Melizabeth, Meow kitties, Merbabu, Mercury, Mercurywoodrose, Mesgul82, Meursault2004, Mgiganteus1, Mic, Micga, Michael Hardy, Michael Zimmermann, Michalws, Mickey gfss2007, Micov, Microtony, Middle Drawer, Mieciu K, Migatu, Mike Rosoft, Miked2144, Mikenorton, Mikewebkist, Mild Bill Hiccup, Militaryboy, Milla297, Mimihitam, Minna Sora no Shita, Minpis, Mirek12, Mirv, Misiek889, Missmarple, Mix321, Miyokan, Miztalkalot, Mjanzen, Mjefm, Mnmazur, Modulatum, Mogism, Molobo, Momo50, Moncrief, Monkeyman, MonoSabio, MoogleEXE, Mortense, Morwen, Mouse-Touch, Mps, Mr. Chainsaw, MrFish, Mrt3366, Msafiri, Mschel, Mushroom, Mvjs, Mx3, Myanw, Mysekurity, Myszodorn, NClement, NHRHS2010, Naddy, Naive cynic, Nakon, NameThatWorks, Naohiro19, Naryathegreat, Nasz, Natalie Erin, Natl1, NawlinWiki, Nayvik, Necronaut, Neferkare, Nein999, Neurorocker, Neverquick, New4325, NewEnglandYankee, NewbieDoo, Ngchen, Niccohel, Niceguyedc, NickW557, Nickfraser, Nicolehere, Nietecza, NigelR, Nightstallion, Nihil novi, Nikkimaria, Niko226, Nima1024, Nite-Sirk, Niteowlneils, Nivix, Njaelkies Lea, Nksd, Nochoje, Nol888, Noozgroop, NorbertArthur, Norm mit, NormanBro, Norum, Notheruser, Notthe600, Nouanoua, NuclearVacuum, NuclearWarfare, Nufy8, Number 57, Numbo3, Nummymuffin, Nv8200p, Nyat, Nyyankees2fn, Oberst, Obli, Oda Mari, Odie5533, Offa of Angel, Ohnder, Ohnoitsjamie, Oktan, Old Moonraker, Oldgregg13, Oleg Alexandrov, Olga Raskolnikova, Olivier, Oliwier1111, Oly Majoieka, Omenge, Omicronpersei8, Omnibus, Oneiros, Oneness01, Only, Ontheglobe, Opelio, Opole.pl, Optimist on the run, Orange5716, Orczar, Oreo Priest, Ottawa4ever, Oxydo, Oxymoron83, P-Chan, PANONIAN, PFHLai, PG-Rated, Pabix, Pacus, Paddyez, Paine Ellsworth, Palosirkka, Pamejudd, Paragon12321, Parhamr, Parrottsvilleelementary, Pat Payne, Patientone, Paul Benjamin Austin, PaulHanson, PaulinSaudi, Paulomazzeirj, PawRon, Pawel bielsko, Pawel z Niepolomic, PawelO, Pawello071, Paweł ze Szczecina, Pawloo, PaxEquilibrium, Pb30, Pedant17, Pel'ican Sh'it, Pensionero, Pernambuko, Peruvianllama, Peter Chastain, Peter Horn, Peter558, PeterisP, Pethr, Petri Krohn, Phaedrus86, Phenz, Phgao, Philip Trueman, Phillip J, Philosopher, PhnomPencil, Picus viridis, Piepiepiepie2, Pilotguy, Pink!Teen, Piotr p928s, Piotrus, Pirawik, Pitdog, Placebo123, Plk, PlushMerga, Pmj, Podruznik, Poeticbent, PolEater, Polacoloco777, Polako, Polaron, PolishMan33, PolishPatriot, Polishtuner9, Politologia, Polski orzel, Polylerus, Porqin, Postdlf, Poszwa, Powertranz, PresN, PrestonH, Programming Hamster, Prusak, Prvc, Przemyśl, Przepla, Pseinstein, Pseudo-Richard, Psy guy, Ptmcd, PuRe$kilz.pl, Pudeo, Pundit, Pwt898, QaviArtilles, Qazwsx13, Qertis, Qogir, Qtoktok, Quadell, Quarma, Quebec99, Quendus, Queson, Quintote, Quizimodo (usurped), Qxz, R Lowry, R9tgokunks, RG2, RGA BALLY, RK, RL0919, Racejr, Radomil, Radzinski, Raficoo, Rajeshm1980, Ran, RapaNui75, Rarelibra, Ravelair, Raven4x4x, Ravenmoon22, Razer rob, RazorICE, Razum2010, Rbarreira, Rd232, Realberserker, RedArmyClub, RedMC, RedWolf, RedWordSmith, ReddShadoe, Redward1, Regulus marzo4103, Regulusmarzo@hotmai.l.com, Rejedef, Rejs12345, Remedios44, Retiono Virginian, Retired username, Rettetast, RexNL, Reywas92, Rfl, Rhys, Rich Farmbrough, Richard Harvey, Richardprins, Richie wright1980, Rick Block, Rickyrab, Rigby27, Rightfully in First Place, Riwnodennyk, Rje, Rjwilmsi, Rklisowski, Rkt2312, Rmhermen, Rmo13, Rmt2m, RoadTrain, Roastytoast, Rob Aleksandrowicz, Rob.derosa, Robenel, Robert Skyhawk, Robert Weemeyer, Robert346, RobertG, Robertgreer, Robertpolska, Robsuper, Roleplayer, Romanm, Romuald Wróblewski, Ronline, Rory096, Roryok, Rowingrocksmysocks, Rreagan007, Rudi Maxer, Ruhrjung, Rumping, Ruok, Ruszewski, Ruud64, Ruy Lopez, Rwerp, RxS, Ryanaxp, Rydia, Ryoho, Ryulong, Ryuu, Rübezahl, SCZenz, SJP, SOPHIAN, SPKirsch, SSJeskimobob, SSZ, SZPANER, Sadi, Sajimem, Salmon L-uh, Salvio giuliano, SamEV, SamWolken, Samah10, Samantha555, Samohyl Jan, Samotny Wędrowiec, Samrolken, Sanbeg, Sango123, Sannse, Sardanaphalus, Satyadasa, Savidan, Sca, Scarletpoet, Scewing, School is fun, Schopenhauer, SchreiberBike, Schwartz und Weiss, Science4sail, Scipius, Scohoust, Scott14, Scoutersig, Scwlong, Sean D Martin, Seb az86556, Sebastian770, Sebek2d, Sebjarod, Secret Saturdays, Secretlondon, Seka'sBreathMints, Selfworm, Sengkang, SenorAnderson, Ser Amantio di Nicolao, Seric2, Serte, SexBloodandMetal, Sfahey, Sfan00 IMG, Sfu, Shalom Alechem, Shanes, Shearonink, Sheitan, Shield, Shoeofdeath, Shoy, Shredder2012, Shuffdog, Shugo255, Sidume, Siemowit, Sietse Snel, Signalhead, Signsolid, Silar, Sillyfolkboy, Simon Peter Hughes, SimonP, Simonxag, Sirius85, Sjwk, Skarl the Drummer, Skatemanman, Skater1123, Skeithkiller, Skier Dude, Skizzik, Skooiman, Skoranka, Skuczera, Skurwiel, Slab stone, Sladen, Slakr, Slawojarek, Slecki19, Sliwers, Smith2006, Smitz, Smokizzy, Smyru, SnappingTurtle, SndrAndrss, Snowfalcon cu, Snowmanradio, Soman, SomeHuman, Someguy1221, Soroush83, Sortior, Soumyasch, South Bay, SpNeo, Space Cadet, Spamtroll, Spicy99, Spitzl, SpookyMulder, Springeragh, SqueakBox, Squids and Chips, Srikeit, Ssolbergj, Stanlavisbad, Starr37, Staszek Lem, Steel, Stefantastic, Stephenb, Steven Weston, Stickee, StoneProphet, Stressbaby 89, Stroganoff, StuMcDee, Subtropical-man, Suna, Sunholm, Sunquanliangxiuhao, Supertask, Supertouch, Surlyduff50, Surowiak, Surtsicna, Suva, Svetovid, SwPawel, SwPawel2, Swedish fusilier, Swid, SwisterTwister, Sycic, Sylwia Ufnalska, Symane, Syvanen, Szater, Szczecin, Szopen, TBadger, TFMcQ, TFOWR, TUF-KAT, TaerkastUA, Tandrasz, Tanessi, Targeman, Tassedethe, Tatry, TaulPaul, Tavrian, Taw, Tawker, Taxman, Tbhotch, Tblack93, Tcc8, TeaDrinker, Team Poland, Teddycom, Tekken50, Template namespace initialisation script, Teodor von Zalenzer Halde, Terfili, Terra Novus, Terrencekeenan, Teysz Kamieński, Th' Duck, Th1rt3en, ThaGrind, Thatdog, Thayts, The Article Creator, The Catholic Knight, The Letter J, The Rambling Man, The sunder king, TheHYPO, TheKMan, TheMaineliner, Theanthrope, Thecoldmidwest, Thehelpfulone, Thelosmets08, Themanwithoutapast, Therequiembellishere, Theroen, Thesoccermage, Thingg, This user has left wikipedia, Thiseye, ThomasK, Thomaspca, Thomehr, Thorsten1, Thryduulf, Thw1309, Tide rolls, TimComm, Timrollpickering, Tiptoety, Titoxd, Tkynerd, Tnick12, Tobby72, Tomaso, Tomasz J Kotarba, Tomirlik, Tommaisey, Tommyjs, Tone, Totti76, TownDown, Tpbradbury, Travelbird, Treisijs, Trekway.com, Trialsanderrors, TrueHistoryPoland, Tryingtomakethingsright, Tsujigiri, Tuckerresearch, Turmerick, Tuspm, Tvieth01, Tweenk, TycoonTim, TygerDawn, Tymek1988, Typobox, Ulmanor, Ulric1313, Ultratomio, Umedard, Underlying lk, UnicornTapestry, Upi, User86654, Usergreatpower, UtherSRG, Uxejn, Valenciano, Valentinian, Valermos, Vampain, Van helsing, VanCleave, Vaquero100, Vardion, VegaDark, Venariul, VenomousConcept, Ventusa, Viewfinder, Viriditas, Visor, Vitund, Vkem, Volunteer Marek, Von Fiszman, VonShroom, Vortexrealm, Vorthax, Vrinan, Vsmith, WGee, WKS Śląsk Wrocław, WOSlinker, WWEDawn, Waldir, Walton One, Wangi, Wavelength, WegianWarrior, Wehberf, WellsSouth, Weorthe, Werdan7, Werideatdusk33, West Brom 4ever, Wgitihm, WhisperToMe, Whomp, Wichid, Wik, Wiki alf, Wikibofh, Wikiborg, Wikidexel2, Wikidudeman, Wikiperuvian, Wilberforcehumphries, Will Beback Auto, William Allen Simpson, William Avery, Wilsonyork, Wimt, Witkacy, Wizzard2k, Wknight94, Woertche.Leo, WojPob, Wojgniew, WolfgangPeters, Wolfrock, Woohookitty, Wtmitchell, Wudkanz, Wulfstan, Wuzzy, Ww2censor, Wyklety, Władysław Komorek, XParadigm777x, XSmartCutie217, Xathaec, Xezbeth, Xlkce, Xtrump, Xx236, YUL89YYZ, Yakudza, Yamaguchi, Yamaka122, Yarovit, Yasis, Yattum, Yatzhek, Ybgursey, YellowMonkey, Yonas29, Yonna, Yorkshirian, Youstill let me make accounts lol, Zabadinho, Zanimum, Zaphod Beeblebrox, Zbihniew, Zello, Zenit4ever, ZimZalaBim, Zinjixmaggir, Zippy, Zlerman, Zoney, Zsinj, Zundark, Zzuuzz, Żlepskż, Δ, Александър, ГДБОБ, 3003 anonymous edits

Image Sources, Licenses and Contributors

file:Poland location map.svg *Source*: http://en.wikipedia.org/w/index.php?title=File:Poland_location_map.svg *License*: unknown *Contributors*: User:NordNordWest

File:Red pog.svg *Source*: http://en.wikipedia.org/w/index.php?title=File:Red_pog.svg *License*: unknown *Contributors*: Anomie

File:Flag of Poland.svg *Source*: http://en.wikipedia.org/w/index.php?title=File:Flag_of_Poland.svg *License*: unknown *Contributors*: Anomie, Mifter

File:Rougon Alpes de Haute Provence France.jpg *Source*: http://en.wikipedia.org/w/index.php?title=File:Rougon_Alpes_de_Haute_Provence_France.jpg *License*: unknown *Contributors*: User:JialiangGao

File:KippelLötschental WoodenHouses.jpg *Source*: http://en.wikipedia.org/w/index.php?title=File:KippelLötschental_WoodenHouses.jpg *License*: unknown *Contributors*: Herzi Pinki

File:Ourika berbere village.jpg *Source*: http://en.wikipedia.org/w/index.php?title=File:Ourika_berbere_village.jpg *License*: unknown *Contributors*: Fathzer / Jean-Marc Astesana from Bretonneux, France

File:Guri Rajasthan 02.jpg *Source*: http://en.wikipedia.org/w/index.php?title=File:Guri_Rajasthan_02.jpg *License*: unknown *Contributors*: User:Entomolo

File:View of RTallavalasa.jpg *Source*: http://en.wikipedia.org/w/index.php?title=File:View_of_RTallavalasa.jpg *License*: unknown *Contributors*: User:Adityamadhav83

File:Ogi Shirakawa02bs3200.jpg *Source*: http://en.wikipedia.org/w/index.php?title=File:Ogi_Shirakawa02bs3200.jpg *License*: unknown *Contributors*: 663highland

File:Pariangan.jpg *Source*: http://en.wikipedia.org/w/index.php?title=File:Pariangan.jpg *License*: unknown *Contributors*: User:MichaelJLowe

File:Kovachevitsa.jpg *Source*: http://en.wikipedia.org/w/index.php?title=File:Kovachevitsa.jpg *License*: unknown *Contributors*: User:Katieleu

File:Village near lake baikal 45230241 f3b18ea67c b.jpg *Source*: http://en.wikipedia.org/w/index.php?title=File:Village_near_lake_baikal_45230241_f3b18ea67c_b.jpg *License*: unknown *Contributors*: Nagyman (flickr)

File:Маяки.jpg *Source*: http://en.wikipedia.org/w/index.php?title=File:Маяки.jpg *License*: unknown *Contributors*: User:ЧуваевНиколай

File:Kosmach Doroha.jpg *Source*: http://en.wikipedia.org/w/index.php?title=File:Kosmach_Doroha.jpg *License*: unknown *Contributors*: User:Yakudza

File:Klenovec.JPG *Source*: http://en.wikipedia.org/w/index.php?title=File:Klenovec.JPG *License*: unknown *Contributors*: User:A1

File:Село Кленовець.JPG *Source*: http://en.wikipedia.org/w/index.php?title=File:Село_Кленовець.JPG *License*: unknown *Contributors*: User:Samuraj18

File:castle.combe.mainstreet.arp.jpg *Source*: http://en.wikipedia.org/w/index.php?title=File:Castle.combe.mainstreet.arp.jpg *License*: unknown *Contributors*: User:Arpingstone

File:Bisley.jpg *Source*: http://en.wikipedia.org/w/index.php?title=File:Bisley.jpg *License*: unknown *Contributors*: User:Jongleur100

File:Saint-Cirq-Lapopie.jpg *Source*: http://en.wikipedia.org/w/index.php?title=File:Saint-Cirq-Lapopie.jpg *License*: unknown *Contributors*: Adam Baker from Houston / Moscow / Toulouse (travel a lot)

File:saifivillage.JPG *Source*: http://en.wikipedia.org/w/index.php?title=File:Saifivillage.JPG *License*: unknown *Contributors*: Elie plus, Linaduliban

File:General view to Al annaze1.jpg *Source*: http://en.wikipedia.org/w/index.php?title=File:General_view_to_Al_annaze1.jpg *License*: unknown *Contributors*: User:Same7same7

File:Puamau.jpg *Source*: http://en.wikipedia.org/w/index.php?title=File:Puamau.jpg *License*: unknown *Contributors*: American at de.wikipedia

File:Carib-fogo-newfoundland-2002.jpg *Source*: http://en.wikipedia.org/w/index.php?title=File:Carib-fogo-newfoundland-2002.jpg *License*: unknown *Contributors*: Conscious, Croquant, EvaK, Peregrine981, Telim tor, WayneRay, 1 anonymous edits

File:Newfane vermont fall 2009.JPG *Source*: http://en.wikipedia.org/w/index.php?title=File:Newfane_vermont_fall_2009.JPG *License*: unknown *Contributors*: chensiyuan

File:A Village in Kaita.jpg *Source*: http://en.wikipedia.org/w/index.php?title=File:A_Village_in_Kaita.jpg *License*: unknown *Contributors*: User:Shamsurabiu

File:POL Krzymów COA.png *Source*: http://en.wikipedia.org/w/index.php?title=File:POL_Krzymów_COA.png *License*: unknown *Contributors*: pl:NN

File:POL powiat koniński gmina Krzymów map.svg *Source*: http://en.wikipedia.org/w/index.php?title=File:POL_powiat_koniński_gmina_Krzymów_map.svg *License*: unknown *Contributors*: user:WarX

File:POL powiat koniński COA.svg *Source*: http://en.wikipedia.org/w/index.php?title=File:POL_powiat_koniński_COA.svg *License*: unknown *Contributors*: user:Bastianow

File:POL województwo wielkopolskie powiat koniński map.svg *Source*: http://en.wikipedia.org/w/index.php?title=File:POL_województwo_wielkopolskie_powiat_koniński_map.svg *License*: unknown *Contributors*: user:WarX

File:POL województwo wielkopolskie flag.svg *Source*: http://en.wikipedia.org/w/index.php?title=File:POL_województwo_wielkopolskie_flag.svg *License*: unknown *Contributors*: Bastianow, Frombenny, Gustavo Szwedowski de Korwin, Ninane, Yarl, 2 anonymous edits

File:Logo Wielkopolskie.png *Source*: http://en.wikipedia.org/w/index.php?title=File:Logo_Wielkopolskie.png *License*: unknown *Contributors*: Bushy moustache, Skier Dude

File:POL województwo wielkopolskie COA.svg *Source*: http://en.wikipedia.org/w/index.php?title=File:POL_województwo_wielkopolskie_COA.svg *License*: unknown *Contributors*: User:Bastianow

File:Wielkopolskie (EE,E NN,N).png *Source*: http://en.wikipedia.org/w/index.php?title=File:Wielkopolskie_(EE,E_NN,N).png *License*: unknown *Contributors*: user:Wulfstan, user:tsca

File:Woj wielkopolskie adm.png *Source*: http://en.wikipedia.org/w/index.php?title=File:Woj_wielkopolskie_adm.png *License*: unknown *Contributors*: MariuszR, Mir, Niki K, Yarl

File:Speaker Icon.svg *Source*: http://en.wikipedia.org/w/index.php?title=File:Speaker_Icon.svg *License*: unknown *Contributors*: Blast, G.Hagedorn, Mobius, Tehdog, 2 anonymous edits

File:Kórnik Zamek 234-07.jpg *Source*: http://en.wikipedia.org/w/index.php?title=File:Kórnik_Zamek_234-07.jpg *License*: unknown *Contributors*: BurgererSF, Madmedea, Pko, Roweromaniak, Sir Gawain, 1 anonymous edits

File:Powstancy Wlkp.jpg *Source*: http://en.wikipedia.org/w/index.php?title=File:Powstancy_Wlkp.jpg *License*: unknown *Contributors*: Masur, Pascal.Tesson, Radomil, Topory

Image:Poznan.StaryRynek.jpg *Source*: http://en.wikipedia.org/w/index.php?title=File:Poznan.StaryRynek.jpg *License*: unknown *Contributors*: User:Łukasz Poznań

File:PL Kalisz Ratusz..JPG *Source*: http://en.wikipedia.org/w/index.php?title=File:PL_Kalisz_Ratusz..JPG *License*: unknown *Contributors*: Bast, Fransvannes, 3 anonymous edits

Image:Gniezno dom.JPG *Source*: http://en.wikipedia.org/w/index.php?title=File:Gniezno_dom.JPG *License*: unknown *Contributors*: User:Südstädter

File:Jezioro Góreckie.jpg *Source*: http://en.wikipedia.org/w/index.php?title=File:Jezioro_Góreckie.jpg *License*: unknown *Contributors*: User:Awersowy

Image:Power plant in Pątnów.jpg *Source*: http://en.wikipedia.org/w/index.php?title=File:Power_plant_in_Pątnów.jpg *License*: unknown *Contributors*: Tomasz Krzykała

Image:Autostrada A2 - Koło.jpg *Source*: http://en.wikipedia.org/w/index.php?title=File:Autostrada_A2_-_Koło.jpg *License*: unknown *Contributors*: Kolanin

File:Polish presidential election 2010, Poznań (10) - Woźniak, Florek.JPG *Source*: http://en.wikipedia.org/w/index.php?title=File:Polish_presidential_election_2010,_Poznań_(10)_-_Woźniak,_Florek.JPG *License*: unknown *Contributors*: User:Yves6

File:Poland administrative division 1999 literki.png *Source*: http://en.wikipedia.org/w/index.php?title=File:Poland_administrative_division_1999_literki.png *License*: unknown *Contributors*: Closedmouth, Dantadd, LMK3, Pere prlpz, Władysław Komorek

File:POL woj dolnoslaskie COA 2009.svg *Source*: http://en.wikipedia.org/w/index.php?title=File:POL_woj_dolnoslaskie_COA_2009.svg *License*: unknown *Contributors*: User:Mboro

File:POL województwo kujawsko-pomorskie COA.svg *Source*: http://en.wikipedia.org/w/index.php?title=File:POL_województwo_kujawsko-pomorskie_COA.svg *License*: unknown *Contributors*: Lech Tadeusz Karczewski

File:POL województwo lubelskie COA.svg *Source*: http://en.wikipedia.org/w/index.php?title=File:POL_województwo_lubelskie_COA.svg *License*: unknown *Contributors*: user:MesserWoland

File:POL województwo lubuskie COA.svg *Source*: http://en.wikipedia.org/w/index.php?title=File:POL_województwo_lubuskie_COA.svg *License*: unknown *Contributors*: User:Bastianow

File:POL województwo łódzkie COA.svg *Source*: http://en.wikipedia.org/w/index.php?title=File:POL_województwo_łódzkie_COA.svg *License*: unknown *Contributors*: User:Bastianow

File:POL województwo małopolskie COA.svg *Source*: http://en.wikipedia.org/w/index.php?title=File:POL_województwo_małopolskie_COA.svg *License*: unknown *Contributors*: user:Poznaniak

File:POL województwo mazowieckie COA.svg *Source*: http://en.wikipedia.org/w/index.php?title=File:POL_województwo_mazowieckie_COA.svg *License*: unknown *Contributors*: User:Poznaniak

File:POL województwo opolskie COA.svg *Source*: http://en.wikipedia.org/w/index.php?title=File:POL_województwo_opolskie_COA.svg *License*: unknown *Contributors*: user:Poznaniak

File:POL województwo podkarpackie COA.svg *Source*: http://en.wikipedia.org/w/index.php?title=File:POL_województwo_podkarpackie_COA.svg *License*: unknown *Contributors*: Derbeth, Frombenny, Gustavo Szwedowski de Korwin, Jaron11, Mir, Valentinian

File:POL województwo podlaskie COA.svg *Source*: http://en.wikipedia.org/w/index.php?title=File:POL_województwo_podlaskie_COA.svg *License*: unknown *Contributors*: Bastianow, Derbeth, Frombenny, Gustavo Szwedowski de Korwin, JDavid, WarX, XtraVert, 2 anonymous edits

File:POL województwo pomorskie COA.svg *Source*: http://en.wikipedia.org/w/index.php?title=File:POL_województwo_pomorskie_COA.svg *License*: unknown *Contributors*: User:Bastianow

File:POL województwo śląskie COA incorrect.svg *Source*: http://en.wikipedia.org/w/index.php?title=File:POL_województwo_śląskie_COA_incorrect.svg *License*: unknown *Contributors*: user:MesserWoland

File:Warminsko-mazurskie herb.svg *Source*: http://en.wikipedia.org/w/index.php?title=File:Warminsko-mazurskie_herb.svg *License*: unknown *Contributors*: Derbeth, Frombenny, Gustavo Szwedowski de Korwin, JDavid, MAC13, Serdelll, 1 anonymous edits

File:POL województwo zachodniopomorskie COA.svg *Source*: http://en.wikipedia.org/w/index.php?title=File:POL_województwo_zachodniopomorskie_COA.svg *License*: unknown *Contributors*: User:Bastianow

File:Poland administrative division 1975 literki.png *Source*: http://en.wikipedia.org/w/index.php?title=File:Poland_administrative_division_1975_literki.png *License*: unknown *Contributors*: J 1982, Ninane, Yarl

File:Poland administrative division 1957 literki.PNG *Source*: http://en.wikipedia.org/w/index.php?title=File:Poland_administrative_division_1957_literki.PNG *License*: unknown *Contributors*: J 1982, Ninane, Yarl

File:Poland administrative division 1922 literki.png *Source*: http://en.wikipedia.org/w/index.php?title=File:Poland_administrative_division_1922_literki.png *License*: unknown *Contributors*: Johnanth, Ninane, Piotrus

File:KingdomofPoland1815.jpg *Source*: http://en.wikipedia.org/w/index.php?title=File:KingdomofPoland1815.jpg *License*: unknown *Contributors*: Mariusz Paździora

File:Rzeczpospolita voivodships.png *Source*: http://en.wikipedia.org/w/index.php?title=File:Rzeczpospolita_voivodships.png *License*: unknown *Contributors*: User:Halibutt

File:Irp.jpg *Source*: http://en.wikipedia.org/w/index.php?title=File:Irp.jpg *License*: unknown *Contributors*: User:Halibutt

File:Flag_of_Poland.svg *Source*: http://en.wikipedia.org/w/index.php?title=File:Flag_of_Poland.svg *License*: unknown *Contributors*: Anomie, Mifter

File:Herb Polski.svg *Source*: http://en.wikipedia.org/w/index.php?title=File:Herb_Polski.svg *License*: unknown *Contributors*: Aotearoa, Denelson83, Frombenny, Gustavo Szwedowski de Korwin, Halibutt, Klemen Kocjancic, Madden, Odder, Zscout370, 3 anonymous edits

File:EU-Poland.svg *Source*: http://en.wikipedia.org/w/index.php?title=File:EU-Poland.svg *License*: unknown *Contributors*: User:NuclearVacuum

File:Increase2.svg *Source*: http://en.wikipedia.org/w/index.php?title=File:Increase2.svg *License*: unknown *Contributors*: Sarang

file:speakerlink-new.svg *Source*: http://en.wikipedia.org/w/index.php?title=File:Speakerlink-new.svg *License*: unknown *Contributors*: User:Kelvinsong

File:Polska 960 - 992.svg *Source*: http://en.wikipedia.org/w/index.php?title=File:Polska_960_-_992.svg *License*: unknown *Contributors*: User:Waćpan, user:Poznaniak

File:Zaprowadzenie chrzescijanstwa 965 Matejko.JPG *Source*: http://en.wikipedia.org/w/index.php?title=File:Zaprowadzenie_chrzescijanstwa_965_Matejko.JPG *License*: unknown *Contributors*: BurgererSF, Goldfritha, Kjetil r, Ludmiła Pilecka, Piotrus, Shakko, Wst, 4 anonymous edits

File:Jan Matejko-Astronomer Copernicus-Conversation with God.jpg *Source*: http://en.wikipedia.org/w/index.php?title=File:Jan_Matejko-Astronomer_Copernicus-Conversation_with_God.jpg *License*: unknown *Contributors*: Alaniaris, BurgererSF, Dirk Hünniger, EugeneZelenko, Gnesener1900, Goldfritha, Kürschner, Ludmiła Pilecka, Matthead, Olivier2, Piotrus, Pko, Plindenbaum, Slomox, Staszek Lem, Wames, Wst, 5 anonymous edits

File:Kraków - Wawel from Kopiec Krakusa.jpg *Source*: http://en.wikipedia.org/w/index.php?title=File:Kraków_-_Wawel_from_Kopiec_Krakusa.jpg *License*: unknown *Contributors*: User:RaNo

File:Rzeczpospolita 1600.png *Source*: http://en.wikipedia.org/w/index.php?title=File:Rzeczpospolita_1600.png *License*: unknown *Contributors*: Bandar Lego, Dcoetzee, Denniss, Electionworld, It Is Me Here

File:SiemiginowskiJerzy.1686.JanIIISobieskiPodWiedniem.jpg *Source*: http://en.wikipedia.org/w/index.php?title=File:SiemiginowskiJerzy.1686.JanIIISobieskiPodWiedniem.jpg *License*: unknown *Contributors*: AnonMoos, Boo-Boo Baroo, BurgererSF, Gryffindor, Infrogmation, Piotrus, Rotational, Shalom Alechem, Umherirrender, Wames, 1 anonymous edits

File:Stanisław II August Poniatowski in coronation clothes.PNG *Source*: http://en.wikipedia.org/w/index.php?title=File:Stanisław_II_August_Poniatowski_in_coronation_clothes.PNG *License*: unknown *Contributors*: Boo-Boo Baroo, BurgererSF, Darwinius, Ecummenic, Kürschner, Mathiasrex

File:Smuglewicz Kosciuszko 2.jpg *Source*: http://en.wikipedia.org/w/index.php?title=File:Smuglewicz_Kosciuszko_2.jpg *License*: unknown *Contributors*: Alma Pater, Bogomolov.PL, Cutwind, DieBuche, DocentX, Goldfritha, Man vyi, MarcusBritish, Mathiasrex, Sfu, Taxiarchos228, Wames, 8 anonymous edits

File:Starcie belwederczykow z kirasjerami rosyjskimi na moscie w Lazienkach.jpg *Source*: http://en.wikipedia.org/w/index.php?title=File:Starcie_belwederczykow_z_kirasjerami_rosyjskimi_na_moscie_w_Lazienkach.jpg *License*: unknown *Contributors*: Allen3, Bedford, Emax, Magog the Ogre, Piotrus, Raven4x4x

File:Jozef Pilsudski.jpg *Source*: http://en.wikipedia.org/w/index.php?title=File:Jozef_Pilsudski.jpg *License*: unknown *Contributors*: Alma Pater, Ejdzej, EugeneZelenko, Kocio, Mathiasrex, Serdelll, 1 anonymous edits

File:Rzeczpospolita 1939.svg *Source*: http://en.wikipedia.org/w/index.php?title=File:Rzeczpospolita_1939.svg *License*: unknown *Contributors*: User:Halibutt

File:10Dyw.JPG *Source*: http://en.wikipedia.org/w/index.php?title=File:10Dyw.JPG *License*: unknown *Contributors*: Andros64, GeorgHH, Jarekt, Mathiasrex, PMG, Peter Weis, Pibwl, Polish29, Robek, Sandmann4u, SuperTank17, Thib Phil, 3 anonymous edits

File:Curzon line en.svg *Source*: http://en.wikipedia.org/w/index.php?title=File:Curzon_line_en.svg *License*: unknown *Contributors*: User:radek.s

File:Polish Soldier's Grave Warsaw 1945.jpg *Source*: http://en.wikipedia.org/w/index.php?title=File:Polish_Soldier's_Grave_Warsaw_1945.jpg *License*: unknown *Contributors*: Tadeusz Bukowski

File:Solidarity poster 1989.jpg *Source*: http://en.wikipedia.org/w/index.php?title=File:Solidarity_poster_1989.jpg *License*: unknown *Contributors*: Andy Dingley, BurgererSF

File:Bundesarchiv Bild 183-F0417-0001-011, Berlin, VII. SED-Parteitag, Eröffnung.jpg *Source*: http://en.wikipedia.org/w/index.php?title=File:Bundesarchiv_Bild_183-F0417-0001-011,_Berlin,_VII._SED-Parteitag,_Eröffnung.jpg *License*: unknown *Contributors*: Fredy.00, Martin H., Mbdortmund, Trust Is All You Need

File:Flaga RP z UE.jpg *Source*: http://en.wikipedia.org/w/index.php?title=File:Flaga_RP_z_UE.jpg *License*: unknown *Contributors*: Michal Osmenda

File:Poland topo.jpg *Source*: http://en.wikipedia.org/w/index.php?title=File:Poland_topo.jpg *License*: unknown *Contributors*: Captain Blood, Hystrix, Juetho, Piotrus

File:Śląskie Kamienie - Dívčí Kameny.JPG *Source*: http://en.wikipedia.org/w/index.php?title=File:Śląskie_Kamienie_-_Dívčí_Kameny.JPG *License*: unknown *Contributors*: User:Pudelek

File:Giewont z Zakopanego.jpg *Source*: http://en.wikipedia.org/w/index.php?title=File:Giewont_z_Zakopanego.jpg *License*: unknown *Contributors*: User:Op

File:2 SPN 01.jpg *Source*: http://en.wikipedia.org/w/index.php?title=File:2_SPN_01.jpg *License*: unknown *Contributors*: BLueFiSH.as, Merlin, Panek, Piotrus, Reytan, Shalom Alechem, Uroboros, 2 anonymous edits

File:Modlin spichlerz.jpg *Source*: http://en.wikipedia.org/w/index.php?title=File:Modlin_spichlerz.jpg *License*: unknown *Contributors*: User:Wojsyl

File:Kurtkowiec i czerwone.jpg *Source*: http://en.wikipedia.org/w/index.php?title=File:Kurtkowiec_i_czerwone.jpg *License*: unknown *Contributors*: Adi, Adziura, Cathy Richards, Daniel Baránek, Dome, Michau Sm, Panek, Selso, Shalom Alechem, ToSter, Ulamm, 5 anonymous edits

File:Solina Poland.jpg *Source*: http://en.wikipedia.org/w/index.php?title=File:Solina_Poland.jpg *License*: unknown *Contributors*: Man vyi, Segu, Shalom Alechem, Silar, 2 anonymous edits

File:Zatoka Pucka - Bay of Puck (8).jpg *Source*: http://en.wikipedia.org/w/index.php?title=File:Zatoka_Pucka_-_Bay_of_Puck_(8).jpg *License*: unknown *Contributors*: Krzysztof, Silar, 1 anonymous edits

File:Biała Ręka Ojcowski PN.jpg *Source*: http://en.wikipedia.org/w/index.php?title=File:Biała_Ręka_Ojcowski_PN.jpg *License*: unknown *Contributors*: Jan Jerszyński

File:WhiteStorkFamily.jpg *Source*: http://en.wikipedia.org/w/index.php?title=File:WhiteStorkFamily.jpg *License*: unknown *Contributors*: User:Manfred Heyde

File:Wisent.jpg *Source*: http://en.wikipedia.org/w/index.php?title=File:Wisent.jpg *License*: unknown *Contributors*: Henryk Kotowski,

File:Plaża w Pucku - kitesurfers - beach in Puck (4).jpg *Source*: http://en.wikipedia.org/w/index.php?title=File:Plaża_w_Pucku_-_kitesurfers_-_beach_in_Puck_(4).jpg *License*: unknown *Contributors*: Fagairolles 34, GiW, Krzysztof, Marek Banach, 3 anonymous edits

File:Bronisław Komorowski rotcropped.jpg *Source*: http://en.wikipedia.org/w/index.php?title=File:Bronisław_Komorowski_rotcropped.jpg *License*: unknown *Contributors*: User:Saibo

File:PL Sejm hall.jpg *Source*: http://en.wikipedia.org/w/index.php?title=File:PL_Sejm_hall.jpg *License*: unknown *Contributors*: User:Network.nt

File:Polishsupremecourt.JPEG *Source*: http://en.wikipedia.org/w/index.php?title=File:Polishsupremecourt.JPEG *License*: unknown *Contributors*: Original uploader was Cjs2111 at en.wikipedia

File:Manuscript of the Constitution of the 3rd May 1791.PNG *Source*: http://en.wikipedia.org/w/index.php?title=File:Manuscript_of_the_Constitution_of_the_3rd_May_1791.PNG *License*: unknown *Contributors*: Ignacy Potocki (1750–1809), Hugo Kołłątaj (1750–1812), Stanisław August Poniatowski (1732–1798)

File:Radoslaw Sikorski meets Secretary Hillary Clinton.jpg *Source*: http://en.wikipedia.org/w/index.php?title=File:Radoslaw_Sikorski_meets_Secretary_Hillary_Clinton.jpg *License*: unknown *Contributors*: Michael Gross, State Department photographer

File:Walesa and Tusk EPP.jpg *Source*: http://en.wikipedia.org/w/index.php?title=File:Walesa_and_Tusk_EPP.jpg *License*: unknown *Contributors*: europeanpeoplesparty

Image:POL location map.svg *Source*: http://en.wikipedia.org/w/index.php?title=File:POL_location_map.svg *License*: unknown *Contributors*: Hiuppo

File:POL województwo dolnośląskie COA.svg *Source*: http://en.wikipedia.org/w/index.php?title=File:POL_województwo_dolnośląskie_COA.svg *License*: unknown *Contributors*: User:Poznaniak

File:POL_województwo_śląskie_COA.svg *Source*: http://en.wikipedia.org/w/index.php?title=File:POL_województwo_śląskie_COA.svg *License*: unknown *Contributors*: user:Poznaniak

File:Krzesiny 8RB.JPG *Source*: http://en.wikipedia.org/w/index.php?title=File:Krzesiny_8RB.JPG *License*: unknown *Contributors*: User:Radomil

File:Polish Army soldiers in Afghanistan.jpg *Source*: http://en.wikipedia.org/w/index.php?title=File:Polish_Army_soldiers_in_Afghanistan.jpg *License*: unknown *Contributors*: Ministerstwo Obrony Narodowej

File:Central Silesian Park - Mounted police 02.jpg *Source*: http://en.wikipedia.org/w/index.php?title=File:Central_Silesian_Park_-_Mounted_police_02.jpg *License*: unknown *Contributors*: User:Yarl

File:Warsaw7ob.jpg *Source*: http://en.wikipedia.org/w/index.php?title=File:Warsaw7ob.jpg *License*: unknown *Contributors*: User:DocentX

File:Schengenzone.svg *Source*: http://en.wikipedia.org/w/index.php?title=File:Schengenzone.svg *License*: unknown *Contributors*: User:CrazyPhunk

File:POL.Gdynia.SeaTowers.DarPomorza2009.jpg *Source*: http://en.wikipedia.org/w/index.php?title=File:POL.Gdynia.SeaTowers.DarPomorza2009.jpg *License*: unknown *Contributors*: User:Wames

File:Gielda3 Wwa beax.jpg *Source*: http://en.wikipedia.org/w/index.php?title=File:Gielda3_Wwa_beax.jpg *License*: unknown *Contributors*: User:Beax

File:Warsaw Downtown.jpg *Source*: http://en.wikipedia.org/w/index.php?title=File:Warsaw_Downtown.jpg *License*: unknown *Contributors*: User:DocenttX

File:Ratusz Zamosc.jpg *Source*: http://en.wikipedia.org/w/index.php?title=File:Ratusz_Zamosc.jpg *License*: unknown *Contributors*: Original uploader was Mceurytos at pl.wikipedia

File:Wrocław, Iglica a Hala století.jpg *Source*: http://en.wikipedia.org/w/index.php?title=File:Wrocław,_Iglica_a_Hala_století.jpg *License*: unknown *Contributors*: Aktron

File:Elektrownia Zeran.jpg *Source*: http://en.wikipedia.org/w/index.php?title=File:Elektrownia_Zeran.jpg *License*: unknown *Contributors*: Original uploader was Hiuppo at pl.wikipedia

File:SP-LRA 2.png *Source*: http://en.wikipedia.org/w/index.php?title=File:SP-LRA_2.png *License*: unknown *Contributors*: © Strower

File:A4 Krakow 20070815 1026.jpg *Source*: http://en.wikipedia.org/w/index.php?title=File:A4_Krakow_20070815_1026.jpg *License*: unknown *Contributors*: User:Jakubhal

File:Marie Curie (Nobel-Chem).png *Source*: http://en.wikipedia.org/w/index.php?title=File:Marie_Curie_(Nobel-Chem).png *License*: unknown *Contributors*: Afrank99, Bohème, DEDB, Fastfission, Kjunix, Martin H., Materialscientist, 3 anonymous edits

File:Warsaw Easter 2008 028.jpg *Source*: http://en.wikipedia.org/w/index.php?title=File:Warsaw_Easter_2008_028.jpg *License*: unknown *Contributors*: Wizzard

File:Poczta Główna Bydg od Brdy a.jpg *Source*: http://en.wikipedia.org/w/index.php?title=File:Poczta_Główna_Bydg_od_Brdy_a.jpg *License*: unknown *Contributors*: User:Pit1233

File:TPSA tower warsaw.JPG *Source*: http://en.wikipedia.org/w/index.php?title=File:TPSA_tower_warsaw.JPG *License*: unknown *Contributors*: User:Masti

File:Gdańsk kamienice przy Długim Targu.jpg *Source*: http://en.wikipedia.org/w/index.php?title=File:Gdańsk_kamienice_przy_Długim_Targu.jpg *License*: unknown *Contributors*: User:Sfu

File:Krakau Markt.jpg *Source*: http://en.wikipedia.org/w/index.php?title=File:Krakau_Markt.jpg *License*: unknown *Contributors*: User:Diether

File:Lichen stary Poland.jpg *Source*: http://en.wikipedia.org/w/index.php?title=File:Lichen_stary_Poland.jpg *License*: unknown *Contributors*: User:Czechu81

File:CerkiewStDuchaBialystok.JPG *Source*: http://en.wikipedia.org/w/index.php?title=File:CerkiewStDuchaBialystok.JPG *License*: unknown *Contributors*: Akir, Ludmiła Pilecka, NVO, Przykuta, Roman Miller, Wouterhagens, Wp, 2 anonymous edits

File:Krakow Synagoga Tempel 20071111 1123 2077.jpg *Source*: http://en.wikipedia.org/w/index.php?title=File:Krakow_Synagoga_Tempel_20071111_1123_2077.jpg *License*: unknown *Contributors*: User:Jakubhal

File:MUG Hospital Bldg 1.jpg *Source*: http://en.wikipedia.org/w/index.php?title=File:MUG_Hospital_Bldg_1.jpg *License*: unknown *Contributors*: User:JohnnyTremolo

File:Collegium Maius 07.JPG *Source*: http://en.wikipedia.org/w/index.php?title=File:Collegium_Maius_07.JPG *License*: unknown *Contributors*: User:Cancre

File:Collegium Minus - UAM.jpg *Source*: http://en.wikipedia.org/w/index.php?title=File:Collegium_Minus_-_UAM.jpg *License*: unknown *Contributors*: User:Mains

File:Ryszard Kaczorowski 21 02 2008 (4).jpg *Source*: http://en.wikipedia.org/w/index.php?title=File:Ryszard_Kaczorowski_21_02_2008_(4).jpg *License*: unknown *Contributors*: Archiwum Kancelarii Prezydenta RP

File:UniwersytetWrocławski.jpg *Source*: http://en.wikipedia.org/w/index.php?title=File:UniwersytetWrocławski.jpg *License*: unknown *Contributors*: User:JacobJ

File:Nikolaus Kopernikus.jpg *Source*: http://en.wikipedia.org/w/index.php?title=File:Nikolaus_Kopernikus.jpg *License*: unknown *Contributors*: ArséniureDeGallium, ArtMechanic, Ausir, Craigboy, Editor at Large, J.delanoy, Kyro, Manuelt15, Matthead, Mikkalai, Pko, Samuel Grant, TarmoK, ThomasPusch, 6 anonymous edits

File:Grand Hotel Sopot.JPG *Source*: http://en.wikipedia.org/w/index.php?title=File:Grand_Hotel_Sopot.JPG *License*: unknown *Contributors*: Alina Zienowicz, e-mail

File:John Paul II Medal of Freedom 2004.jpg *Source*: http://en.wikipedia.org/w/index.php?title=File:John_Paul_II_Medal_of_Freedom_2004.jpg *License*: unknown *Contributors*: Eric Draper

File:Frederic Chopin photo.jpeg *Source*: http://en.wikipedia.org/w/index.php?title=File:Frederic_Chopin_photo.jpeg *License*: unknown *Contributors*: Louis-Auguste Bisson, very old and poor copy, completely restored and remastered by Amano1

File:Tomasz Stanko-CP.jpg *Source*: http://en.wikipedia.org/w/index.php?title=File:Tomasz_Stanko-CP.jpg *License*: unknown *Contributors*: Cezary Piwowarski

File:Moderne wolkenkrabber Warschau 0875.PNG *Source*: http://en.wikipedia.org/w/index.php?title=File:Moderne_wolkenkrabber_Warschau_0875.PNG *License*: unknown *Contributors*: User:VanNouten

File:Fakty TVP-studio Wroclaw 2008.jpg *Source*: http://en.wikipedia.org/w/index.php?title=File:Fakty_TVP-studio_Wroclaw_2008.jpg *License*: unknown *Contributors*: Original uploader was SNOL at pl.wikipedia

File:Adam Mickiewicz.PNG *Source*: http://en.wikipedia.org/w/index.php?title=File:Adam_Mickiewicz.PNG *License*: unknown *Contributors*: Arek1979, Bohème, BurgererSF, Butko, Darwinius, Freta, Kilom691, Krzysiu Jarzyna, Mathiasrex, Mattes, Rocket000, Shakko, Thuresson, TwoWings, 2 anonymous edits

File:Henryk Sienkiewicz.PNG *Source*: http://en.wikipedia.org/w/index.php?title=File:Henryk_Sienkiewicz.PNG *License*: unknown *Contributors*: Kazimierz Mordasewicz

File:Władysław Reymont.jpg *Source*: http://en.wikipedia.org/w/index.php?title=File:Władysław_Reymont.jpg *License*: unknown *Contributors*: Andros64, Demut, HuBar, 2 anonymous edits

File:Witold Gombrowicz by Bohdan Paczowski - detail.jpg *Source*: http://en.wikipedia.org/w/index.php?title=File:Witold_Gombrowicz_by_Bohdan_Paczowski_-_detail.jpg *License*: unknown *Contributors*: Bohdan_Paczowski

File:Czeslaw Milosz, 1986.jpg *Source*: http://en.wikipedia.org/w/index.php?title=File:Czeslaw_Milosz,_1986.jpg *License*: unknown *Contributors*: User:MDCarchives

File:Ratusz2007.jpg *Source*: http://en.wikipedia.org/w/index.php?title=File:Ratusz2007.jpg *License*: unknown *Contributors*: adamchemik

File:Kościół parafialny p.w. Wniebowzięcia NMP (Mariacki), Kraków, Rynek Główny, A-3 01.jpg *Source*: http://en.wikipedia.org/w/index.php?title=File:Kościół_parafialny_p.w._Wniebowzięcia_NMP_(Mariacki),_Kraków,_Rynek_Główny,_A-3_01.jpg *License*: unknown *Contributors*: User:Piotrekwas

File:Bigos in a cafe at Wawel Castle.jpg *Source*: http://en.wikipedia.org/w/index.php?title=File:Bigos_in_a_cafe_at_Wawel_Castle.jpg *License*: unknown *Contributors*: Colin Cameron from Edinburgh, Scotland

File:National Stadium in Warsaw, Poland - Greece.jpg *Source*: http://en.wikipedia.org/w/index.php?title=File:National_Stadium_in_Warsaw,_Poland_-_Greece.jpg *License*: unknown *Contributors*: grotos

File:Openstreetmap logo.svg *Source*: http://en.wikipedia.org/w/index.php?title=File:Openstreetmap_logo.svg *License*: unknown *Contributors*: OpenStreetMap

MIX
Papier aus verantwortungsvollen Quellen
Paper from responsible sources
FSC® C105338

Printed by Books on Demand GmbH, Norderstedt / Germany